Guide to Broadcasting Stations

Newnes Technical Books

The Butterworth Group

United Kingdom	Butterworth & Co (Publishers) Ltd London: 88 Kingsway, WC2B 6AB
Australia	Butterworths Pty Ltd Sydney: 586 Pacific Highway, Chatswood, NSW 2067 Also at Melbourne, Brisbane, Adelaide and Perth
Canada	Butterworth & Co (Canada) Ltd Toronto: 2265 Midland Avenue, Scarborough, Ontario, M1P 4S1
New Zealand	Butterworths of New Zealand Ltd Wellington: T & W Young Building, 77—85 Customhouse Quay, 1, CPO Box 472
South Africa	Butterworth & Co (South Africa) (Pty) Ltd Durban: 152—154 Gale Street
USA	Butterworth (Publishers) Inc Boston: 10 Tower Office Park, Woburn, Mass. 01801

First published 1946
Eighteenth edition 1980 by Newnes Technical Books

© Butterworth & Co (Publishers) Ltd, 1980

British Library Cataloguing in Publication Data

Guide to broadcasting stations. — 18th ed.
  1. Radio stations — Directories
  I. 'Wireless world' guide to broadcasting
  stations
  621.3841'6'025   TK6555

  ISBN 0 408 00467 3

Typeset by Scribe Design, Gillingham, Kent
Printed in England by Redwood Burn Ltd.,
Trowbridge and Esher

CONTENTS

ACKNOWLEDGEMENT

Thanks are due to the BBC for the lists of broadcasting stations, which were prepared by the BBC Receiving Station, Caversham Park, Reading.

A GUIDE TO LISTENING

AERIALS

For medium- and long-wave reception most receivers have an internal ferrite-rod aerial, which enables them to receive the local stations and the stronger of the more distant stations. These aerials are directional and give very poor results when the rod points in the direction of the transmitter, so it is worthwhile checking whether the aerial is favourably oriented. Some portable receivers have a turntable built into the base to enable them to be rotated conveniently, and larger receivers sometimes have a control which rotates the aerial within the case. In searching the wavebands, it is easily possible to miss signals from transmitters in line with the aerial, and it is a good plan, therefore, to repeat the search with the aerial at right angles to its former position. Ferrite-rod aerials are not used for short-wave reception and these directional effects are not present.

Many receivers have aerial and earth sockets and it is possible to effect a great improvement in reception by using an external aerial. Suitable forms of aerial are discussed later. When an external aerial is used the effect on reception of rotating the ferrite rod is much less marked and may even be absent altogether.

Short-wave receivers often have telescopic aerials which can be extended to a metre or so in length and can sometimes be tilted. These, too, can provide satisfactory reception of the stronger signals.

Improved reception is often possible using an aerial external to the receiver, supported, for example, on the wall of a room or in the roof-space. Results from indoor aerials are, however, often disappointing because the aerial is screened from the wanted signals by the walls and/ or roof of the building and is near the electrical wiring and domestic electrical equipment. While it may be easy to suppress noise and interference from your own washing machine and light dimmer, it is less easy to suppress your neighbour's, which in flats and terraces may be even nearer than your own. Indoor aerials are thus liable to pick up a high level of electrical interference.

For best results an outdoor aerial is essential and, if electrical interference is a problem, the aerial should be located in an interference-free area and special precautions taken to ensure that the cable connecting the aerial to the receiver does not pick up interference from the electrical system of the house.

An inverted-L aerial, Fig. 1a, is quite suitable for long- and medium-wave reception. Results improve as the length of the horizontal section and the height above the ground are increased. The horizontal section should be insulated from the supporting wires or ropes by several small porcelain insulators at each end. The downlead should be a continuous length of wire with the aerial and not joined separately because soldered and other kinds of joints are likely to deteriorate with weathering and eventually cause crackles and other effects in the receiver. The lead-in should be arranged to drop from the aerial well away from the building to avoid contact with gutters and to minimize pick-up of noise from the domestic electrical supply. If a tree is used to support the far end of the

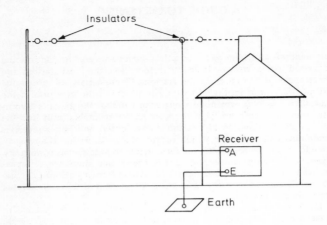

Fig. I(a). Inverted-L aerial ...

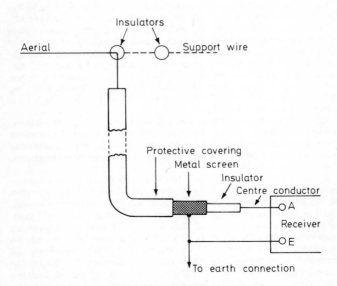

... (b) and screened down-lead

2

aerial, allowance must be made for the movement of the tree under windy conditions. The terminating wire or rope should be passed over a pulley and terminated with a suitable weight. In this way the tension in the aerial wire can be maintained independent of movement of the tree.

Sometimes it is convenient to take the downlead from the centre point of the horizontal section. The resulting aerial is known as a T-aerial and its performance is very similar to that of the inverted-L.

As a precaution against electrical interference the downlead can take the form of a coaxial cable, the inner conductor providing the connection to the receiver and the outer conductor being earthed as shown in Fig. 1b. By this means the downlead is screened so that only signals picked up by the horizontal wire are conveyed to the receiver.

Where there is insufficient space for an inverted-L or T-aerial or where electrical interference is a serious problem, a vertical rod say 5 m long may be used. This should be mounted in an area where interference is a minimum (a chimney top is often a suitable place) and connected to the receiver by a screened lead as shown in Fig. 2. Aerial manufacturers market kits containing all the parts for such an installation including matching transformers for use at the aerial base and receiver input.

An inverted-L, T-aerial or vertical rod aerial is suitable for short-wave reception but where space permits there are more efficient types which can be used: these are directional aerials which should therefore

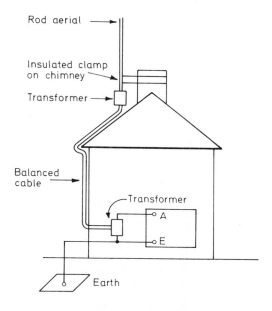

Fig. 2. Vertical rod aerial

3

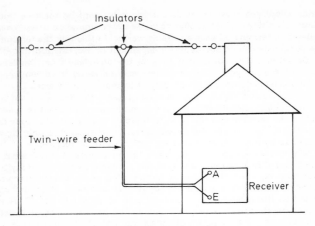

Fig. 3. Simple dipole aerial

be positioned to favour the direction of the transmitters it is desired to receive.

One suitable aerial is the half-wave dipole illustrated in Fig. 3. It consists of two horizontal arms connected to the receiver by a balanced feeder. The dipole should be mounted as high as possible but 10 m is probably the maximum height which is convenient for most domestic situations. The length of each of the two horizontal arms should be chosen to suit the wavelength of the signals it is desired to pick up and varies between 13 m for the 49-m band to 3 m for the 11-m band. The aerial has maximum response to signals travelling at right angles to its length and has minimum response to transmissions arriving in line with the aerial.

A disadvantage of the simple dipole is that it is less effective on wavebands other than those for which it has been designed. If, however, the two leads of the feeder are connected together and to the receiver aerial terminal, the earth terminal being connected to ground, the aerial then becomes a T type which can be used for long- and medium-wave reception as well as for short waves. A two-pole change-over switch can be used to convert the aerial from the dipole to the T form.

A better form of directional short-wave aerial is the inverted-V, Fig. 4. This provides a greater signal to the receiver than the simple dipole and by using the dimensions shown it can be effective over all the short-wave bands. It requires only a single support pole, one end of the aerial being earthed via a 400-ohm terminating resistor, the other being connected to the receiver input. This aerial has maximum sensitivity to signals travelling in the plane of the aerial as indicated in the diagram.

The Beverage aerial demands length but not height and consists of a length of wire supported by a series of short poles, say 2 or 3 m high and spaced sufficiently close to prevent undue sag. Each should be surmounted with an insulator to which the wire is bound, not looped, the

4

aerial being terminated at the far end by a 600-ohm resistor. Wire length is not critical but it should not be less than about 50 m and the lead-in should be direct to the receiver without significant deviation from the general line; if this can be achieved an r.f. transformer and coaxial line are not required to connect the aerial to the receiver. This aerial favours the reception of signals travelling in line with the aerial from the terminating resistor end, and is used professionally with wire lengths up to 1000 m.

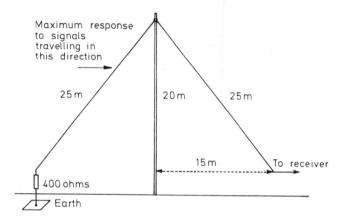

Fig. 4. Inverted-V aerial

When a receiver is supplied from a 3-pole main socket there is a natural temptation to use the earthed pole of the socket as an earth connection for the receiver. Such a connection is likely to be unsatisfactory because the physical connection of the main earth to ground is often at a considerable distance from the mains socket. Consequently the earth path may have appreciable resistance and can carry signals capable of causing interference to radio reception.

Where a receiver is provided with a signal earth terminal, local interference may be reduced by connecting the terminal by a short lead to a copper plate or earth rod buried in the ground. A similar connection is also required for inverted-V and some other aerials. A connection to a gas pipe is usually an unsatisfactory earth and may be extremely dangerous; most underground metal gas pipes are being replaced by plastics pipes. A connection to a metal water pipe is satisfactory only if the pipe is connected directly to an underground water main: in many modern housing estates the metal pipes within the house are connected to buried polythene pipes and do not provide a satisfactory earth connection.

Propagation of radio waves is a complex subject and in this brief chapter we can give only a general description of those aspects which may interest those whose hobby is listening to broadcasts generally and

who may be sufficiently enthusiastic to extend their listening to more distant and difficult signals.

A knowledge of the basic facts will ensure that listening is carried out at the right time of day for a given frequency and will certainly provide more enjoyment by enabling the listener to anticipate good reception conditions and eliminate fruitless searching when propagation is poor. Awareness of the trends in propagation will leave the listener in no doubt as to causes of changes in reception and will enable selection of the most favourable periods for searching for the weaker and seldom-heard signal.

There are good reasons why a particular broadcast may within a short period improve to a degree when programme content can be appreciated or conversely may virtually disappear. It can also happen that strong signals from a given area may suddenly disappear within a minute or two, yet are received at their former strength thirty minutes or more later. Normal fading of signals may become more rapid, accompanied by a fall in strength and a corresponding increase in noise. These are some of the effects which the listener will observe and which, if carefully considered, will enable assessment of some of the changes in the ionosphere which affect reception conditions.

The basic facts governing short-wave propagation can be summarized in the following way. Short-wave radio communication is achieved by waves which strike the ionosphere (electrified layers in the earth's upper atmosphere) at an oblique angle and are reflected back to earth to cover the receiving area. The waves may be reflected again when they strike the earth and reach other receiving areas after successive bounces from the ionosphere. However in certain areas, for example in the area between the transmitter and the first earth-reflection point, the transmission may be very difficult to receive: this is a so-called skip zone.

For satisfactory short-wave communication the frequency must be chosen with care. If it is too high, the waves penetrate the ionosphere and are lost in space: if it is too low the waves are attenuated by absorption in the lower regions of the ionosphere. Best results are achieved by using the highest frequency which does not penetrate the ionosphere and the value of this, the highest probable frequency (HPF), depends on the degree of ionization of the gases in the ionosphere. This in turn depends largely on the extent to which the ionosphere over the chosen path is illuminated by the sun. Thus the HPF varies with the time of day and with the time of year.

Any changes in the degree of ionization of the reflecting layer can affect long-distance reception and such changes can be produced by increased radiation from the sun, e.g. from blemishes on its surface such as sunspots and invisible areas called M regions. As seen from the earth, the sun takes 27 days to rotate on its axis and some effects on reception, particularly those due to long-lived M regions, tend to have a 27-day periodicity. Moreover the incidence of sunspots follows an 11-year cycle; this in turn causes an 11-year periodicity in short-wave reception conditions.

At any particular time, a survey of all the broadcast bands will indicate that some are very active (many stations being receivable, possibly with a fair amount of interference), while other bands may appear to be

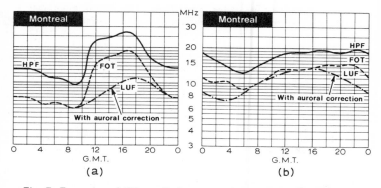

Fig. 5. Examples of HF prediction curves for the U.K.-Montreal path for January (a) and July (b). The highest probable frequency (HPF) is the median usable frequency exceeded on 10% of the days. The LUF (lowest usable frequency) curves are for commercial telegraphy and assume the use of high-power transmitters and rhombic aerials. The path to Montreal passes through the Northern Auroral Zone and waves are subject to additional absorption: a correction is made for this in calculating the LUF. The term optimum traffic frequency (FOT) is self explanatory

practically devoid of signals, apart from weak scattered radiation from stations some few hundred kilometres from the receiving site. These situations arise because transmissions are so arranged that programmes can be received at maximum signal strength in a desired area at local peak listening time. The choice is governed largely by HPF applicable to the required ionospheric path at that time, but the precise frequency may be somewhat lower to ensure that day-to-day variations in HPF do not seriously affect reception throughout the period of the programme or of the transmission schedule, which may be required to continue without alteration for a number of months. Two examples of prediction curves are given in Fig. 5. The upper curve represents the HPF and, in general, frequencies above this value are heard infrequently. The lower curve indicates the frequency below which the signal-to-noise ratio of the received signal becomes unacceptable. If frequencies between these two boundary curves are used the transmitted wave normally propagates over the particular path and provides a service in the target zone. Frequencies which approach the HPF produce the stronger signal but their propagation is more likely to be affected by ionospheric disturbances. It is impossible to predict with accuracy the variations to which signals are likely to be subjected, although short-term predictions based on daily observation of signals received can provide fair accuracy.

It is not good practice to make frequent changes of frequency in a broadcast schedule because the listener expects to find the programme at the same spot on the tuning scale. Thus to offset the variations of

MUF and make best use of the transmission paths, two or more transmitters are used to radiate the same programmes on different frequencies. Thus a programme may be radiated simultaneously on say the 17, 15 and possibly the 11 MHz bands, so that when the HPF is high the 17 MHz signal is good and well supported by 15 MHz, whilst the low-frequency channel may suffer from some absorption. When the HPF is low, the 17 MHz signal is weak and a better service is obtained on 15 and 11 MHz.

Announcements made prior to close-down and radiated by all broadcasts in the same network mention the frequency of the broadcast band which is closing and that which is opening. For any target zone the peak listening time is evening and the schedules of tranmissions to that area are arranged to provide programmes at that time. Frequency separation on the short-wave bands is only 5 kHz and there may be difficulty in receiving a programme clear of interference.

The broadcast bands and their frequency limits are shown elsewhere in this book, and in general transmissions must, by international agreement, be confined to these bands. Other services are similarly restricted to certain frequencies. The highest allotted frequency used in short-wave broadcasting is 26.100 MHz: thus when the HPF exceeds that figure, maximum use of propagation conditions cannot be obtained. However, most domestic receivers have an upper tuning limit as low as 21 or even 17 MHz.

Comparison of Fig. 5a and b shows that under summer-time conditions the HPF curves flatten considerably, day-time frequencies being lower and night-time frequencies higher than in winter-time. In the summer more transmissions are crowded into fewer bands and interference problems increase.

At periods of minimum solar activity HPF's are generally lower throughout the year and the reduced spectrum available for broadcasting causes increased interference.

Sunspot maximum conditions occur in 1979 and there will be a gradual decrease in the HPFs until sunspot minimum conditions are reached in 1985, after which the HPFs will increase toward the next maximum.

The ionosphere is subject to disturbances which can affect radio reception. The disturbances are usually caused by sunspots and their effect is to make the reception of certain of the short-wave broadcast bands difficult or even impossible. Thus, under certain conditions, signals in the high-frequency bands may be weak although the low-frequency bands are normal. Alternatively, the high-frequency bands may be normal and the low-frequency bands weak. Under more exceptional circumstances all the broadcast bands may be inaudible.

Thus, if short-wave reception is found to be very poor, the most likely cause is a disturbance in the ionosphere and it is unlikely to last more than a few days. Most of the disturbances last only a few hours.

SIGNAL IDENTIFICATION

Tuning scales of receivers are often marked with a wealth of station names, but it does not follow that all these stations can be received, even with a good external aerial. Equally, it should not be assumed that

stations, even if they can be received, will be picked up at precisely the point indicated by the name on the scale. The calibration of a receiver is not always exact, even when it is new, and it tends to drift as the receiver gets older. Calibration can be checked by tuning in certain stations which maintain their allotted frequencies with great accuracy. Most transmitters have a reasonably good frequency stability but those on 200 KHz, 5, 10, 15 and 20 MHz are particularly accurate. For further details of standard frequencies and time codes consult the Wireless World Diary.

Signal identification involves a knowledge of broadcasting organizations and their programmes, transmission schedules and target areas, rather than merely a knowledge of transmitting stations. Interval signals, clock chimes, times of operation, types of programme and signal strength also aid identification.

The large number of languages used in short-wave broadcasting would be beyond the ability of one person to learn, but consistent listening to broadcasts from known countries, many radiating similar versions of the current world news, gives good practice in recognizing languages. The sound pattern of an unrecognized language can be compared with other broadcasts of languages which appear similar, remembering that a dialect may be used. Knowledge of the normal occupants of a waveband in terms of broadcasters and their programme schedules is also useful in language recognition.

Interval signals, or particular tunes, are often used to preface the start of transmissions or programmes, typical examples being the use of Bow Bells, Greenwich Time Signal and Big Ben by the BBC, the Canadian National Anthem by Sackville. the Kremlin Bells by Moscow and the Kookaburra by Melbourne. Eastern European stations often use the first few bars of a well-known melody, which may have been written by an eminent composer.

If these signals can be recorded on tape, a library of interval signals can be built up. Each recording can be annotated with the details of reception, to increase its usefulness as a reference guide.

The make-up and timing of broadcasts can often prove useful in identification. If a continuous programme is well balanced between music, speech, drama and other items, it is probably intended for home consumption and the opening and closing times of the transmission will give some idea of the time of day in the country of origin. A programme consisting of short items, with a preponderance of speech, starting or finishing at odd times, is likely to be a service for listeners outside the country. Clock chimes may narrow the choice, by fixing the time zone, and they often precede an announcement or news bulletin. Don't forget that some countries have summer or daylight saving time. The relaying of programmes can produce difficulties; for instance, London's Big Ben is heard from stations all over the world. Nevertheless, continued listening may provide a clue, which can be a change of atmosphere at the conclusion of a relay, or an announcement that follows.

Most broadcasts begin with a period of tone for technical alignment purposes, followed by an interval signal and announcement, then possibly a time check, and finally the programme. The frequency of the

line-up tone differs from one organization to another; thus the BBC uses 1 kHz, Federal Germany 900 Hz, and some authorities use 440 Hz.

The close-down of a transmission is also important, because of the probability of announcements, and perhaps a national anthem or clock chime.

The type of programme may yield evidence of the nationality of the broadcasting organization and of the intended zone of reception. Domestic services can generally be recognized by the parochial nature of the news, the coverage of world events being small. Programmes for a country's nationals abroad are often a blend of domestic and world news, with commentaries in the national language; a typical example is the BBC World Service. Frequent news bulletins, almost exclusively concerned with world events and given in many languages, strongly suggest a service intended for foreign listeners.

When a programme whose source is unknown is sufficiently intelligible to be followed to a limited extent and a guess made at the language, a search for the identical programme on different frequencies may help identification. A second receiver is useful for this, because it can be tuned to known stations operating services in the supposed language. If another transmission carrying the programme is found, it may be assumed that both originate from the same source, though not necessarily from co-sited transmitters. One transmission may be a relay, and if so the quality of the unknown transmission may not be as good as the known.

It may still be difficult to determine the location of the unknown station, though listening at times of programme change for local or regional announcements can help in reaching a conclusion. At such times there may be changes in fading characteristics and background noise, indicating the conclusion of a relay and suggesting that the signal has been affected twice by ionospheric conditions. A typical example of relays is provided by the BBC World Service broadcast from the UK and relayed by bases in the Middle East, Far East and South Atlantic; other examples are provided by Deutsche Welle in Germany and its relay base in Africa, by Paris and Brazzaville, and by the Voice of America at Greenville and its overseas stations at Tangier, Munich, Monrovia and elsewhere.

The stronger of two signals carrying the same programme may not necessarily be that of the nearer transmitter. The receiving location may be in the skip zone of this transmitter and thus obtains a weaker signal. A better signal may also be obtained from the more distant transmitters if this is beamed toward the receiver site.

Programmes which are broadcast simultaneously on a fair number of frequencies can be generally quickly identified as belonging to the same country or programme network. Even if foreign languages cause difficulty, the sound pattern of any language may indicate that the programme is originating from the same source irrespective of the number of transmitter outlets it may be heard on. With some experience, it becomes possible to identify languages without understanding them; thus, if Cairo broadcasting in Arabic is positively identified, it is then feasible to recognize Arabic programmes in the external service of another country.

If a simultaneous broadcast cannot be found, but the programme pattern can be established, a search of programme schedules issued by the various countries may show details which conform closely to those of the unknown station.

A tape recorder is useful to aid identification, to give positive proof of reception, and to provide a tape library of announcements and call signs, and the interval signals and jingles which characterize so many programmes and broadcast services. The tape machine should be close to the receiver and available for immediate use with its input connected to the receiver output, the mains supply switched on and a tape ready to record.

Any announcement heard which is not readily identifiable may be recorded and later played back repeatedly to help in identifying the language or recognizing some feature. Microphone facilities are useful to enable details of the time, date and approximate frequency or wavelength to be added to the recorded announcement. Such recordings could well form the beginning of an index of station announcements, which might later be arranged in country or geographical order to facilitate further research.

Tape recordings can be made of the signature tunes which most stations use either prior to their opening announcement or before particular programmes. Signature tunes are usually repeated for some minutes before the scheduled opening time, and as indicated previously, they may consist of a well-known melody characteristic of the country, of a few tones, or of bells or clock chimes. These tunes, when memorized, can provide an instant means of identification, but while some are distinctive, others are not, and a tape recording is often useful for comparison.

RECEPTION REPORTS

Reports on reception are always welcomed by broadcasting organizations, whether the listener is located in the target area or not. Such reports can provide useful information on transmissions, and help the broadcaster to assess the accuracy of the assessments on which his schedule was based and the effectiveness of the service.

Reception reports should be concise and accurate and should follow established form. This is preferable to a letter, which takes time to read and assess, and may require the extraction and tabulation of detail by qualified engineers to make it suitable for comparison with other similar reports. The assimilation of reports in a large broadcasting organization must follow a procedure requiring minimum effort, and this is possible only if listeners set out their reports in a standard manner. The information given can then be quickly and accurately assessed by junior staff, who may be trained to present the results in a form suitable for analysis by computer.

The detail which can be provided in a reception report is, however, quite large, and is of great importance when it is based on a test transmission. Information on every aspect of such transmissions is required, and each reception report is studied in detail. Where broadcasts follow a

pattern or schedule of long standing, much detail can be omitted and the report can be shortened. The analysis of abbreviated reports of daily reception conditions supplies the transmission schedule engineer with a constant flow of information on signal strength, interference and overall merit. Thus any deviation from normal reception is easily detectable and can be investigated. Possibly the ionospheric path may have changed and a different frequency or aerial array may be needed; perhaps new interference has appeared and steps must be taken to eliminate or avoid it.

SINPFEMO, SINPO AND SIO

The generally recognized form for reports is based on the SINPFEMO code. Each letter signifies a particular aspect of reception and is followed by a rating figure (1 to 5) the significance of which is indicated in the table.

SINPFEMO code

Symbol and Meaning		1	2	3	4	5
S	Signal strength	barely audible	poor	fair	good	excellent
I	Interference	extreme	severe	moderate	slight	nil
N	Noise	extreme	severe	moderate	slight	nil
P	Propagation disturbance	extreme	severe	moderate	slight	nil
F	Frequency of fading	very fast	fast	moderate	slow	nil
E	Modulation quality	very poor	poor	fair	good	excellent
M	Modulation depth	over-mod.	poor/nil	fair	good	maximum
O	Overall merit	unusable	poor	fair	good	excellent

Restricted forms of this code are now more commonly used, for example, SINPO, in which no indication is given of the frequency of fading or the quality and depth of modulation. An even simpler code is SIO, which embraces only three criteria, namely signal strength, interference and overall merit. The number of rating figures has also been reduced: this is possible because if a signal is classified as 1 reception is unusable, and the difference bwteen 4 and 5 is so small in short-wave reception that the higher of these can be ignored. Where signals are poor enough to justify a rating of less than 2, or where interference is non-existent, 0 may be used.

Reception report forms are available from most broadcasting organizations on request from listeners who indicate their willingness to provide reports on a continuing basis, and some notes on the compilation of a SINPO report are given below. A full SINPFEMO report could be provided merely by adding the F, E and M criteria.

The use of the code is simple if care is taken in assessing the value of the signal. Few broadcasts other than those from a local transmitter qualify for rating of S5 or O5, but with these exceptions all other ratings are feasible. Enthusiasm should not be allowed to distort the report and the signal should be analysed with some precision for each aspect of the SINPO code.

The strength of the signal reported on can be compared with that of well-known broadcasts and the assessment is even simpler if the receiver has a tuning meter indicating signal strength. Such meters are often calibrated in dB above one microvolt, but the calibration is frequently incorrect and should not be accepted unless means are available of checking it.

The assessment of interference depends on the type and character of the interfering signal. This signal is often a whistle or heterodyne note, caused by reception of two signals with a carrier-frequency difference less than the bandwidth of the receiver. Thus, if the receiver bandwidth is 8 or 9 kHz, and the interfering signal is say 3 or 4 kHz from the wanted broadcast, a heterodyne whistle of this frequency is audible. The interference is, however, more troublesome if the frequency difference is only 1 to 2 kHz, because the ear is more sensitive at these lower frequencies. Even though the strength of two interfering signals may be the same, if one is displaced 4 kHz and the other 1 kHz from the wanted signal, a rating of 14 may apply to one and of 13 to the other. Similarly, a weak background of programme is less disturbing than a whistle or steady tone. Thus the rating to be entered is a measure of the intelligibility of the wanted signal.

Atmospheric noise is seldom worse than N3, except during periods of ionospheric disturbance, summer static or the precipitation of electrified rain. Good conditions, and the use of the higher frequency bands, do not normally produce ratings better than N4. Care should be taken to ignore noise introduced by the receiver, especially when the signal is weak and the receiver is operating at full gain.

Propagation disturbance may be more difficult to assess: it is related to the intensity of atmospheric noise and the degree of fading of the received signal. If noise is high and fading rapid, but the programme can be followed, a rating of P3 is justified, but rapid fading to a depth causing programme mutilation qualifies for P1. If little or no noise is apparent and the fades are shallow and do not exceed about one second, being well held by a.g.c., the rating should be P4 or P5.

Overall merit is assessed by taking the average of the individual rating figures to the nearest whole number. There is no need to add a plus or minus sign, or to indicate small differences in merit, because each rating in the code is intended to cover a wide range of conditions and if the listener is certain that one rating does not apply the next figure must be correct.

The best report loses its value if the listener fails to give such essential details as his name and address, the date and time of reception, and the approximate frequency or wavelength (the waveband alone is not enough). Any definitely identified interference should be specified, but if this cannot be done, details of the type of programme or other interfering signal should be mentioned.

LONG AND MEDIUM-WAVE EUROPEAN STATIONS

This list includes only those stations which are believed to be active on the frequencies indicated and which may be heard in Europe. Certain stations located outside the Continent of Europe are sometimes heard in Western Europe and these are included in this section, although they are situated outside the 'European Broadcasting Area', as defined in the Copenhagen Plan. This area is bounded on the south by 30° north latitude, that is, by the territories bordering the Mediterranean Sea, excluding those parts of Arabia and Saudi Arabia within this area but including Iraq. On the west it encloses Iceland, Eire and the Azores, and on the east it is bounded by the meridian 40° east of Greenwich.

Stations are listed against the frequency on which they have been heard, which may in some cases be the frequency allocated in the Copenhagen Plan. Wavelength in metres is shown beside the frequency. The power is in kW.

Alternative station names or exact location of transmitters, where known, are shown after the usual station name. In appropriate cases station names have been given the anglicized spelling.

In certain instances, groups of low powered stations are indicated by a numeral following the name of the main station in the group, e.g. Cagliari + 5, the figure being the number of additional stations to that named.

Abbreviations used in the list are as follows:—

AFN	American Forces Network
BBC	British Broadcasting Corporation
BRT	Belgische Radio en Televisie (in Dutch)
COPE	Cadena de Ondas Populares Espanolas
DDR	Deutscher Demokratischer Rundfunk
IBA	Independent Broadcasting Authority
NDR	Norddeutscher Rundfunk
Prog	Programme
R	Radio
RCE	Radiocadena Espanola
RIAS	Rundfunk im Amerikan Sektor von Berlin
RNE	Radio Nacional de Espana
RTBF	Radiodiffusion-Television Belge (in French)
SER	Sociedad Espanola de Radiodifusion
St	Station
VOA	Voice of America

kHz	Metres	Station	Country	Power	Programme
155	1936	Brasov	Roumania	1200	1st Programme
		Donebach	Germany (W)	250	Deutschlandfunk
		Tromsoe	Norway	10	
		Engels	USSR	150	
164	1829	Allouis	France	2000	France-Inter
		Tachkent	USSR	150	
173	1734	Kaliningrad	USSR	1000	
		Lvov	Ukraine	500	
182	1648	Saarlouis-Felsberg	Germany (W)	2000	Europe No. 1
		Ankara	Turkey	1200	2nd Programme
		Oranienburg-Rehmate	Germany (E)	750	Stimme Der DDR
		Alma Ata	USSR	250	
191	1571	Motala	Sweden	300	1st Programme
		Caltanissetta	Italy	10	National Programme
		Tbilissi	USSR	500	
200	1500	Droitwich	UK	400	Radio 4 UK
		Burghead	UK	50	Radio 4 UK
		Westerglen	UK	50	Radio 4 UK
		Warszawa	Poland	200	2nd Programme
		Etimesgut	Turkey	200	1st Programme
		Leningrad	USSR	150	
		Moscow	USSR	100	
209	1436	Azilal	Morocco	800	Prog A (Arabic)
		Kiev	Ukraine	500	
		Reykjavik + 1	Iceland	100	
		Mainflingen	Germany	15	Deutschlandfunk
218	1376	Monte-Carlo	Monaco	1400	
		Oslo	Norway	200	1st Programme
		Baku + 1	USSR	500	
227	1322	Warszawa 1	Poland	2000	1st Programme
236	1271	Junglinster	Luxembourg	2000	
		Kichinev + 2	USSR	1000	
245	1224	Kalundborg	Denmark	200	1st Programme
		Taldy Kurgan + 3	USSR	500	
		Erzurum	Turkey	200	1st Programme
254	1181	Tipaza	Algeria	1500	French Programme
		Lahti	Finland	200	Programme 1
		Duchanbe + 4	USSR	300	
263	1141	Moskva	USSR	2000	
		Burg	Germany (E)	200	
		Irkutsk + 1	USSR	1000	
272	1103	Ceskoslovensko	Czechoslovakia	1500	Hvezda Programme Czech/Slovak
		Novosibirsk	USSR	150	
281	1068	Minsk	Bielorussia	500	
		Achkabad + 1	USSR	150	

kHz	Metres	Station	Country	Power	Programme
433		Oulu	Finland	10	Programme 1
520	577	Innsbruck Aldrans + 2	Austria	10	
		Hof-Saale + 1	Germany (W)	0.2	Bayerischer Rundfunk
529	567	Ain Beida	Algeria	600	Arabic Programme
531	565	Beromunster	Switzerland	500	German Programme
		Jerusalem	Israel	200	
		Greifswald	Germany (E)	500	Radio DDR 1
		Torshavn	Denmark	5	
		Cheboksary	USSR	30	
540	555.5	Carraroe	Ireland	2	
		Solt	Hungary	2000	1st Programme
		Tripoli	Libya	600	Arabic Programme
		Waver-Overijse	Belgium	150	
		Oulu 1	Finland	10	Programme 1
		Sulaibiyah	Kuwait	1500	
		Orenburg	USSR	100	
549	546	Les Trembles	Algeria	600	Arabic Programme
		Kichinev + 4	USSR	1000	
		Nordkirchen + 1	Germany (W)	100	Deutschlandfunk
		Beli Kriz	Yugoslavia	6	
558	537	Abu-Zaabal	Egypt	40	
		Faro	Portugal	10	1st Programme
		Targu Jiu + 3	Roumania	200	2nd Programme
		Helsinki 1	Finland	100	Programme 1
		Monte-Ceneri Passo	Switzerland	100	Italian Programme
		Maribor	Yugoslavia	100/20	
		Rostock + 1	Germany (E)	20	DDR 1
		Qeslagh	Iran	1000	
567	529	Homs	Syria	300	General Service Arabic
		Valenca do Minho	Portugal	10	1st Programme
		Tullamore	Ireland	500	R Telefis Eireann
		Berlin	Germany (W)	100	Sender Freies Berlin
		Brasov + 1	Roumania	50	
		Bologna + 3	Italy	25	National Programme
		Volgograd	USSR	250	
576	521	Bechar	Algeria	400	Arabic Programme
		Braganca	Portugal	1	1st Programme
		Vidin	Bulgaria	1000	Sofia 2
		Riga + 1	USSR	500	
		Stuttgart	Germany (W)	300	Suddeutscher Rundfunk
		Schwerin	Germany (E)	250	DDR 1
		Tel-Aviv 2	Israel	200	

kHz	Metres	Station	Country	Power	Programme
585	513	Gafsa	Tunisia	350	National Programme Arabic
		Wien-Bisamberg + 3	Austria	600/240	
		Madrid	Spain	200	RNE 1
		Paris + 1	France	10	France-Inter Paris et Provence
		Riyadh	Saudi Arabia	1200	
		Perm	USSR	30	
594	505	Oujda 1	Morocco	100	Prog A (Arabic)
		Frankfurt + 1	Germany (W)	400	Hessischer Rundfunk
		Pleven	Bulgaria	250	Sofia 1
		Ijevsk	USSR	150	
603	497.5	Lyon-Tramoyes	France	300	France Culture
		Oradea + 2	Roumania	50	
		Koenigswuster-hausen	Germany (E)	20	DDR 1
		Nicosia	Cyprus	20	
		Newcastle	UK	2	Radio 4 UK
		Mariehamn	Finland	1	Swedish Programme
		Tiumen	USSR		
612	490	Sarajevo + 2	Yugoslavia	600	
		Sebaa Aioun	Morocco	300	Prog A (Arabic)
		Tallinn + 1	USSR	100	
		Tullamore	Ireland	200	
		Qashr'Shirin	Iran	400	
621	483	Vila Real + 1	Portugal	10	1st Programme
		Batra	Egypt	450	
		Wavre-Overijse	Belgium	150	RTBF 1 (French)
		Ukhta	USSR	150	
		Santa Cruz Tenerife	Canary Is.	100	RNE 1
630	476	Tunis-Djedeida	Tunisia	600	National Programme Arabic
		Miranda do Douro	Portugal	1	1st Programme
		Cukurova	Turkey	300	1st Programme
		Timisoara	Roumania	400	2nd Programme
		Dannenberg	Germany (W)	10	Sender Freies Berlin
		Vigra	Norway	100	1st Programme
		Saratov	USSR		
639	469	Praha	Czechoslovakia	1500	Praha Prog (Czech)
		La Coruna + 3	Spain	100	RNE 1
		Zakaki	Cyprus	100	BBC
		Bonab	Iran	400	
		Omsk	USSR		
648	463	Crowborough	UK	500	External Service

kHz	Metres	Station	Country	Power	Programme
648 contd	463	Orfordness	UK	50	
		Simferopol + 1	Ukraine	150	
		Plovdiv	Bulgaria	30	2nd Programme
		Murska Sobota	Yugoslavia	10	
657	457	Tel-Aviv 1	Israel	200	
		Napoli + 4	Italy	120	National Programme
		Murmansk	USSR	150	
		Tchernovtsy	Ukraine	25	
		Neubranden-burg	Germany (E)	20	Berliner Rundfunk
		Tantan	Morocco	50	Prog A (Arabic)
666	450	Damas-Sabboura	Syria	100	General Service Arabic
		Vilnius + 1	USSR	500	
		Bodensee-sender	Germany (W)	300/180	Sudwestfunk
		Lisboa	Portugal	135	1st Programme
		Sombor	Yugoslavia	10	
675	444	Marseille	France	600	France-Inter
		Benghazi	Libya	100	
		Jerusalem	Israel	20	
		Lopik	Netherlands	120	3rd Programme
		Ujgorod + 1	Ukraine	50	
		Bodoe	Norway	10	1st Programme
684	438.5	Beograd	Yugoslavia	1000	
		Sevilla	Spain	250	RNE 1
		Hof-Saale	Germany (E)	100	RIAS
		Mashad	Iran	100	
		Tselinograd	USSR	300	
693	433	Basra	Iraq	1200	
		Viseu	Portugal	10	1st Programme
		Berlin	Germany (E)	250	Berliner - Rundfunk
		Barrow	UK	1	Radio 2
		Bexhill	UK	1	Radio 2
		Brighton	UK	1	Radio 2
		Burghead	UK	50	Radio 2
		Droitwich	UK	150	Radio 2
		Exeter	UK	1	Radio 2
		Folkestone	UK	1	Radio 2
		Plymouth	UK	1	Radio 2
		Postwick	UK	10	Radio 2
		Redmoss	UK	1	Radio 2
		Stagshaw	UK	50	Radio 2
		Nicosia	Cyprus	20	1st Programme
		Ufa	Israel	150	
702	427	Banska Bystrica + &	Czechoslovakia	400	Bratislava (Slovak) + Regional
		Monte-Carlo	Monaco	300	2nd Programme
		Radio Andorra	Andorra	250	R. Andorra

18

kHz	Metres	Station	Country	Power	Programme
702 contd	427	Umraniye	Turkey	150	2nd Programme
		Sebaa-Aioun	Morocco	140	Prog C (Berber)
		Finnmark	Norway	20	1st Programme
		Aachen-Stolberg + 3	Germany (W)	5/1,2	NDR/WDR
		Duchanbe + 1	USSR	50	
711	422	Jerusalem	Israel	10	
		Rennes 1	France	300	France Culture
		Donetsk	Ukraine	150	
		Abu-Zaabal	Egypt	100	
		Tallinn + 3	USSR	50	
		Heidelberg + 4	Germany (W)	5	Suddeutscher Rundfunk
		Nis	Yugoslavia	20	
720	417	Norte 1	Portugal	100	
		Sfax	Tunisia	200	Regional Programme
		Holzkirchen	Germany (W)	150	Radio Free Europe
		Zakaki	Cyprus	100	BBC
		Lisnagarvey	UK	10	Radio 4
		London	UK	500	Radio 4
		Londonderry	UK	0.25	Radio 4
		Bailesti + 10	Roumania	2	
		Taybad	Iran		
729	411.5	Mirandela	Portugal	10	1st Programme
		Athinai	Greece	150	National Programme
		Oviedo + 3	Spain	50	RNE 1
		Puttbus	Germany (E)	10	DDR 1
		Sadiyat	United Arab Emirates	750	
738	406.5	Tel-Aviv 2	Israel	1200	
		Poznan	Poland	300	2nd Programme
		Barcelona	Spain	250	RNE 1
		Tcheliabinsk	USSR	150	
747	402	Sarakeb	Syria	100	General Service (Arabic)
		Petrich	Bulgaria	500	Sofia 2
		Lopik	Netherlands	250	2nd Programme
		Sarajevo + 1	Yugoslavia		
		Ouagadougou	Upper Volta	100	
		Karaganda	USSR		
756	397	Lugoj + 1	Roumania	1000	
		Braunschweig	Germany (W)	800/200	Deutschlandfunk
		Salman Pack	Iraq	300	
		Lisboa	Portugal	135	
		Kuopio	Finland	10	Programme 1
		Redruth	UK	2	Radio 4 S West
		Carlisle	UK	1	
		Radio Malta	Malta	20	
765	392	Sottens	Switzerland	500	French Programme

19

kHz	Metres	Station	Country	Power	Programme
765 contd	392	Odessa	Ukraine	150	.
		Medvejiegorsk	USSR	150	
		Ioannina	Greece	20	Armed forces
		Dakar	Senegal	400	
774	388	Abis	Egypt	1000	
		Voronej + 2	USSR	150	
		Caceres + 3	Spain	60	RNE 1
		Stockholm	Sweden	100	1st Programme
		Split + 3	Yugoslavia	50	
		Sofia + 1	Bulgaria	30	Sofia 1
		Klagenfurt + 10	Austria	30	4th Programme
		Leeds	UK	0.5	Local Radio
		Farahabad	Iran	20	
783	383	Tartus	Syria	600	
		Miramar	Portugal	100	3rd Programme
		Burg	Germany (E)	1000	Stimme Der DDR
		Kiev + 2	Ukraine	100	
		Kazan	USSR		
792	379	Kavalla	Greece	500	VOA
		Limoges 1	France	300	France Culture
		Sevilla EAJ.5	Spain	20	SER
		Jesenik + 2	Czechoslovakia	1	
		Astrakhan	USSR	50	
801	374.5	Amman	Jordan	200	
		Leningrad	USSR	1000	
		Munchen-Ismaning + 1	Germany (W)		Bayerischer Rundfunk
		Barnstaple	UK	2	Radio 4 S West
		Ulan-Ude + 3	USSR	1000	
810	370	Skopje	Yugoslavia	1000	
		Burghead	UK	100	
		Dumfries	UK	2	Radio Scotland
		Redmoss	UK	5	
		Westerglen	UK	100	
		Madrid EAJ.7	Spain	20	SER
		Berlin 3	Germany (W)	5	BBC
		Abu Dhabi	United Arab Emirates	50	
		Volgograd	USSR	150	
819	366	Sud Radio	Andorra	600	Sud Radio
		Batra 2	Egypt	450	
		Warszawa	Poland	300	2nd Programme
		Trieste	Italy	25	National Programme
		Rabat 1	Morocco	25	Prog A (Arabic)
828	362	Sebha	Libya	300	
		Oujda 2	Morocco	100	Prog B (French)
		Shumen + 1	Bulgaria	500	Sofia 2
		Hannover + 1	Germany (W)	100	NDR/WDR
		Freiburg	Germany (W)	40	Sudwestfunk

kHz	Metres	Station	Country	Power	Programme
828 contd	362	Barcelona EAJ.1	Spain	20	
		Castelo Branco	Portugal	1	1st Programme
		Gorkii	USSR	150	
837	358	Beyrouth	Lebanon	100	
		Ponta Delgada	Azores	10	
		Nancy 1	France	200	France Culture
		Kharkov	Ukraine	150	
		Novi Sad + 1	Yugoslavia	50	
		Las Palmas EAK.35	Canary Is.	10	COPE
846	355	Zefat	Israel	5	
		Roma	Italy	540	2nd Programme
		Moskva	USSR	60	
		Ceske Bude- jovice + 1	Czechoslovakia	30	Prague Prog (Czech)
		Elista	USSR	30	
855	351	Bucuresti	Roumania	750	2nd Programme
		Murcia + 2	Spain	125	RNE 1
		Berlin	Germany (W)	100	RIAS
		Plymouth	UK	1	Radio 4 S West
		Blackburn	UK	0.5	Local Radio
		Amman	Jordan	10	
		Tchelhbinsk	USSR	150	
864	347	Ksar es Souk	Morocco	15	Prog A (Arabic)
		Paris	France	300	France Culture
		Plovdiv	Bulgaria	150	Sofia 1
		Erevan	USSR	150	
873	344	Leningrad	USSR	150	
		Frankfurt	Germany (W)	150	AFN
		Abu Zaabal	Egypt	50	
		Zaragoza EAJ.101	Spain	20	SER
		Budapest + 1	Hungary	20	2nd Programme
		Gijon EAJ.34	Spain		
882	340	Titograd	Yugoslavia	100	
		Koenigswuster- hausen	Germany (E)	250	
		Washford	UK	70	Radio Wales
		Penmon	UK	10	Radio Wales
		Tywyn	UK	5	Radio Wales
		Wrexham	UK	2	Radio Wales
		Damman	Saudi Arabia		
		Stavropol	USSR		
891	337	Antalya	Turkey	600	1st Programme
		Ujgorod + 1	Ukraine	150	
		Hengelo- Overijssel	Netherlands	10	Relay Lopik 2 & 3
		Alger	Algeria	200	

kHz	Metres	Station	Country	Power	Programme
900	333	Milano	Italy	600	National Programme
		Brno + 4	Czechoslovakia	30	
		Iochkar-Ola + 2	USSR	50	
909	330	Thourah	Iraq	200	
		Bournemouth	UK	1	Radio 2
		Brookmans Park	UK	140	Radio 2
		Clevedon	UK	50	Radio 2
		Fareham	UK	1	Radio 2
		Lisnagarvey	UK	10	Radio 2
		Londonderry	UK	1	Radio 2
		Moorside Edge	UK	100	Radio 2
		Redruth	UK	2	Radio 2
		Torquay	UK	1	Radio 2
		Westerglen	UK	50	Radio 2
		Whitehaven	UK	1	Radio 2
		Cluj + 2	Roumania	50	2nd Programme
		Iman	USSR	50	
918	327	Paphos	Cyprus	50/2	
		Ljubljana	Yugoslavia	600	
		Madrid EAJ.29	Spain	20	
		Mezen	USSR	100	
927	324	Wolvertem	Belgium	300	BRT 1 (Dutch)
		Izmir	Turkey	200	1st Programme
		Zakynthos	Greece	50	Regional Programme
		Lamego + 1	Portugal	1	3rd Programme
		Nebit Dag	USSR	50	
936	320.5	Cairo	Egypt	100	
		Agadir 1	Morocco	600	Prog A (Arabic)
		Bremen + 1	Germany (W)	100	Radio Bremen
		Lvov	Ukraine	500	
		Djakovica + 1	Yugoslavia	10	
		Rio de Janeiro	Brazil	100	
		Rezaieh	Iran	10	
945	317.5	Toulouse 1	France	300	France Culture
		Rostov/Don + 1	USSR	300	
		Pleven	Bulgaria	30	2nd Programme
		Miercurea Ciuc	Roumania	14	2nd Programme
		Larissa	Greece	10	Armed forces
		Volgograd	USSR		
954	314.5	Brno + 3	Czechoslovakia	750	Prague (Czech) & Regional
		Trabzon	Turkey	300	1st Programme
		Madrid EAJ.2	Spain	20	RCE
		Iraklion	Greece	20	Armed forces
		Al Arish	Qatar	750	
963	311.5	Tunis	Tunisia	100	
		Beograd	Yugoslavia	200	
		Sofia	Bulgaria	150	Sofia 1

kHz	Metres	Station	Country	Power	Programme
963	311.5	Turku	Finland	100	Programme 1
contd		Paris 4	France	8	Radio Sorbonne
		Coimbra + 1	Portugal	1	
		Soba	Sudan	200	
972	309	Marrakech 1	Morocco	1	Prog A (Arabic)
		Nikolaev	Ukraine	500	
		Hamburg + 2	Germany (W)	300	NDR/WDR
		Korca	Albania	15	
981	306	Alger	Algeria	200	
		Megara	Greece	200	Armed forces
		Goeteborg	Sweden	150	1st Programme
		Trieste	Italy	10	Regional Service
		Bor + 2	Yugoslavia	10	
		Cheb + 2	Czechoslovakia	1	
		Shiraz	Iran	400	
990	303	Amchit	Lebanon	100	
		Berlin-West + 1	Germany (W)	300	RIAS
		Barcelona EAJ.15	Spain	10	SER
		Exeter	UK	1	Radio 4 S West
		Wolverhampton	UK	0.3	IBA Beacon Radio
		Shiraz	Iran	400	
999	300	Kukes	Albania	8	
		Kichinevq	USSR	100	
		Hoyerswerda + 2	Germany (E)	20	Berliner Rundfunk
		Rimini + 1	Italy	6	Second Programme
		Delimara	Malta	5	
		Fareham	UK	0.3	Local Radio S West (BBC)
		Nottingham	UK	1.0	IBA Radio Trent
		Tselinograd	USSR	30	
1008	298	Kerkyra	Greece	50	Regional Programme
		Lopik	Netherlands	300	1st Programme
		Aleksinac	Yugoslavia	200	
		Mozyr + 3	Bielorussia	50	
		Asswan	Egypt	1	
		Semman	Iran	20	
		Las Palmas EAJ.50	Canary Is.	10	SER
1017	295	Tetuan	Morocco	20	
		Istanbul	Turkey	1200	1st Programme
		Wolfsheim	Germany (W)	600	Sudwestfunk
		Nitra + 4	Czechoslovakia	30	Hvezda & Regional
		Venezia + 1	Italy	25	Regional Service
		Tripolis	Greece	10	Armed forces
1026	292	Jerusalem	Israel	200	
		Graz-Dobl + 14	Austria	100	

kHz	Metres	Station	Country	Power	Programme
1026	292	Brest + 2	Bielorussia	5	
contd		Rabat	Morocco	1	Prog B (French)
		Safi	Morocco	1	Prog A (Arabic)
		Vigo EAJ.48	Spain	10	SER
		Madrid	Spain		
		Alicante EAJ.31	Spain		
		Belfast	UK	1	IBA Downtown Radio
		Tabriz	Iran	100	
1035	290	Porto Alto	Portugal	120	
		Babylon	Iraq	2000	
		Tallinn	USSR	500	
		Milano + 9	Italy	50	
		Gillingham	UK	1	Local Radio
		Sheffield	UK	1	Local Radio
		Yazd	Iran	20	
1044	287	Thessaloniki	Greece	50	Regional Programme
		Burg	Germany (E)	1000	DDR 1
		Sebaa Aioun	Morocco	300	Prog B (French)
		Tbilissi	USSR	200	
1053	285	Tripoli	Libya	50	
		Tetuan 2	Morocco	20	Prog C (Berber)
		Iasi	Roumania	1000	
		Barnstaple	UK	1	Radio 1
		Barrow	UK	1	Radio 1
		Bexhill	UK	2	Radio 1
		Brighton	UK	2	Radio 1
		Burghead	UK	20	Radio 1
		Droitwich	UK	150	Radio 1
		Dundee	UK	1	Radio 1
		Folkestone	UK	1	Radio 1
		Hull	UK	1	Radio 1
		Londonderry	UK	1	Radio 1
		Postwick	UK	10	Radio 1
		Stagshaw	UK	50	Radio 1
		Start Point	UK	100	
		Kurgan	USSR	50	
1062	282	Norte	Portugal	100	2nd Programme
		Diyabakir	Turkey	300	1st Programme
		Kalundborg	Denmark	100	3rd Programme
		Cairo	Egypt	50	
		Cagliari + 6	Italy	25	National Programme
		Zagreb + 1	Yugoslavia	10	
		Saransk	USSR	150	
1071	280	Tartus	Syria	60	
		Brest + 4	France	100	France-Inter
		Riga + 3	USSR	60	
		Banja Luka	Yugoslavia	25	

kHz	Metres	Station	Country	Power	Programme
1071 contd	280	Mnich Hradiste + 1	Czechoslovakia	25	
		Dniepro- petrovsk	Ukraine	20	
		Mesolongion	Greece		
		Rajkot + 1	India	1000	
1080	278	Katowice	Poland	1500	
		Koper-Belikriz	Yugoslavia	200	
		Mallorca EAJ 3	Spain	5	SER
		Granada EAJ 16	Spain	2	SER
		Casablanca	Morocco	1	Prog A (Arabic)
		Crestias	Greece	20	Armed forces
		Abadan	Iran	600	
1089	275.5	Krasnodar	USSR	300	
		Brookmans Park	UK	150	Radio 1
		Fareham	UK	1	Radio 1
		Lisnagarvey	UK	10	Radio 1
		Moorside Edge	UK	150	Radio 1
		Redmoss	UK	2	Radio 1
		Redruth	UK	2	Radio 1
		Tywyn	UK	1	Radio 1
		Washford	UK	50	Radio 1
		Westerglen	UK	50	Radio 1
		Whitehaven	UK	1	Radio 1
		Durres	Albania	100	
		Novi-Sad	Yugoslavia	20	
1098	273	Bratislava	Czechoslovakia	400	Bratislava (Slovak) & Reg.
		Bologna	Italy	60	Regional Service
		Alma-Ata + 1	USSR	150	
		Santa Cruz de Palma	Canary Is.	5	
1107	271	Batra	Egypt	600	
		Kaunas + 4	USSR	150	
		Novi-Sad	Yugoslavia	150	
		Munchen + 4	Germany (W)	100	AFN
		Madrid EFE.14	Spain	20	RCE
		Valladolid EFE.1	Spain	2	RCE
1116	269	Ouarzazate + 1	Morocco	15	Prog A (Arabic)
		Bari + 5	Italy	150	2nd Programme
		Kaliningrad	USSR	30	
		Derby	UK	0.5	Local Radio
1125	267	El Beida	Libya	500	
		Stara Zagora	Bulgaria	500	Sofia 2
		Zagreb + 1	Yugoslavia	200	

kHz	Metres	Station	Country	Power	Programme
1125	267	La Louviere	Belgium	150	RTBF 2 (French)
contd		Leningrad	USSR	20	
1134	264.5	Zagreb/ Tovarnik + 3	Yugoslavia	300	
		Sevilla EAK.2	Spain	10	COPE
		Bilbao EAK.13	Spain	10	COPE
		Zaragoza EAK.6	Spain	10	COPE
		Valencia EAK.5	Spain	10	COPE
		San Sebastian EAK.44	Spain	2	COPE
		Zamora	Spain		
		Calcutta	India	1000	
		Sulaibiyah	Kuwait	750	
1143	262	Kaliningrad	USSR	150	
		Nova Gradiska + 1	Yugoslavia	100	
		Les Trembles	Algeria	40	Prog A (Arabic)
		Stuttgart + 13	Germany (W)	10	AFN
		Messina	Italy	6	2nd Programme
		Duchanbe + 1	USSR	150	
1152	260	Cluj 2 + 1	Roumania	950	2nd Programme
		Birmingham	UK	1	IBA BRMB Radio
		Glasgow	UK	2	IBA R Clyde
		London	UK	5.5	IBA LBC
		Manchester	UK	1	IBA Piccadilly Radio
		Plymouth	UK	0.5	IBA Plymouth Sound
		Tyneside	UK	1	IBA Metro Radio
		Marrakech	Morocco	1	Prog B (French)
		Tabriz	Iran	100	
1161	258	Strasbourg + 2	France	200	France-Inter
		Tanta	Egypt	200	
		Biala Slatina + 1	Bulgaria	150	Sofia 1
		Duchanbe	USSR	1000	
1170	256	Porto	Portugal	10	
		Moghilev	Bielorussia	1000	
		Beli Kriz + 1	Yugoslavia	50	
		Ipswich	UK	0.5	IBA R Orwell
		Portsmouth	UK	0.8	IBA R Victory
		Swansea	UK	0.8	IBA Swansea Sound
		Teeside	UK	1.0	IBA R Tees
		Erfurt + 1	Germany (E)		Berliner Rundfunk
1179	254	Bacau + 2	Roumania	200	
		Thessaloniki	Greece	50	Forces Radio
		Hoerby	Sweden	100	1st Programme
		Murcia	Spain	5	SER

kHz	Metres	Station	Country	Power	Programme
1179	254	Badalona E	Spain	5	SER
contd		EAJ.39			
1188	252.5	Szolnok + 1	Hungary	135	2nd Programme
		Cork 1	Ireland	10	R Telefis Eireann
		San Remo	Italy	6	
		Kuurne	Belgium	5	BRT 2 (Dutch)
		Casablanca	Morocco	1	Prog B (French)
		Teheran	Iran	100	
1197	251	Portalegre	Portugal	1	1st Programme
		Munchen-Ismaning	Germany (W)	300	VOA
		Minsk + 3	Bielorussia	50	
		Agadir	Morocco	20	Prog C (Berber)
		Alexandria	Egypt	10	
		Cambridge	UK	0.1	Radio 3
		Sana'a	Yemen	10	
1206	249	Bordeaux 1	France	100	France Culture
		Wroclaw + 3	Poland	200	
		Haifa	Israel	10	
1215	247	Durres	Albania	240	
		Brighton	UK	1	Radio 3
		Brookmans Park	UK	50	Radio 3
		Burghead	UK	20	Radio 3
		Droitwich	UK	50	Radio 3
		Fareham	UK	1	Radio 3
		Hull	UK	0.15	Radio 3
		Lisnagarvey	UK	10	Radio 3
		Londonderry	UK	0.25	Radio 3
		Moorside Edge	UK	50	Radio 3
		Newcastle	UK	2	Radio 3
		Plymouth	UK	1	Radio 3
		Postwick	UK	1	Radio 3
		Redmoss	UK	1	Radio 3
		Redruth	UK	2	Radio 3
		Tywyn	UK	0.5	Radio 3
		Washford	UK	60	Radio 3
		Westerglen	UK	40	Radio 3
		Tartu + 2	USSR	50	
		Las Palmas ECS 4	Canary Is.	20	
1224	245	Vidin	Bulgaria	1000	Sofia 2
		Beer-Sheva	Israel	10	Forces Station
		Madrid EAK 1	Spain	20	COPE
		Villarreal EAK 78	Spain	2	COPE
		Rio de Janeiro	Brazil	100	
		Djambul	USSR	150	
1233	243	Praha + 4	Czechoslovakia	750	Hvezda & Regional

kHz Metres	Station	Country	Power	Programme
1233 243	Cape Greco	Cyprus	600	
contd	Tanger	Morocco	200	R Tanger
	Liege	Belgium	5	RTBF 2 (French)
1242 241.5	Marseille 2	France	150	France Culture
	Kiev + 3	Ukraine	150	
	Vaasa	Finland	25	Swedish Programme
	Tel Aviv	Israel	10	
	Muscat	Oman	100	
1251 240	Tripoli	Libya	500	Arabic Programme
	Chaves	Portugal	1	1st Programme
	Siofok + 1	Hungary	135	2nd Programme
	Amsterdam	Netherlands	1	Relay Lopik 3
	Dublin	Ireland	20	R Telefis Eireann
1260 238	Rhodos	Greece	500	VOA
	Szczecin	Poland	160	2nd Programme
	San Sebastian EAJ 8	Spain	10	SER
	Valencia EAJ 3	Spain	20	SER
	Alcala Henares	Spain	5	
1269 236	Neumunster	Germany (W)	600	Deutschlandfunk
	Novi-Sad	Yugoslavia	100	
	Radio Paradise	British West Indies	50	
1278 235	Strasbourg	France	300	France Culture
	Odessa	Ukraine	150	
	Florina	Greece	20	Forces Programme
	Turku 2	Finland	4	Swedish Programme
	Bradford	UK	0.1	IBA Pennine Radio
	Assuit	Egypt	10	
	Kabul	Afghanistan	100	
1287 233	Tel-Aviv 1	Israel	100	Forces Station
	Lisboa	Portugal	2.5	
	Ceskoslovensko 2 + 3	Czechoslovakia	300	Hvezda & Regional
	Mytilini	Greece	5	Forces Programme
	Lar	Iran	20	
1296 231	Orfordness	UK	500	External Services
	Loznica + 1	Yugoslavia	10	
	Rabat	Morocco	1	Prog C (Berber)
	Semnan	Iran	10	
	Rio de Janeiro	Brazil		
	Sennar	Sudan	1500	
	Duchanbe + 1	USSR	1000	
1305 230	Constantine 2 + 1	Algeria	40	French Network
	Haifa	Israel	20	
	Rzeszow + 3	Poland	100	
	Marche	Belgium	50	RTBF 2 (French)
1314 228	Shkoder	Albania	10	
	Stavanger	Norway	100	1st Programme

kHz	Metres	Station	Country	Power	Programme
1314	228	Constantza + 2	Roumania	50	1 & 2 Programme
contd		Tripolis	Greece	20	Forces Programme
		Ancona + 3	Italy	6	2nd Programme
		Shkoder	Albania	10	
		Valencia	Spain	18	RCE
		Madrid ECS.11	Spain	20	RCE
		Cadiz EJF.5	Spain	2	RCE
		Skopje + 1	Yugoslavia	20	
		Fort de France	Martinique	50	
		Arbadil	Iran	20	
1323	227	Leipzig	Germany (E)	150	Soviet Programme
		Zyyi	Cyprus	50	BBC
		Targu Mures + 3	Roumania	15	
1332	225	Funchal	Madeira	10	
		Elvas	Portugal	1	1st Programme
		Roma + 3	Italy	300	National Programme
		Parnu + 2	USSR	30	
		Brno Mesto + 4	Czechoslovakia	25	
		Gnjilane + 3	Yugoslavia	2	
		Teheran	Iran	100	
1341	224	Lisnagarvey	UK	100	Radio Ulster
		Londonderry	UK	0.25	Radio Ulster
		Budapest-Lakihegy	Hungary	300	2nd Programme
		Delimara	Malta	20	
		Alma-Ata + 1	USSR	30	
		Alexandria + 1	Egypt	10	
1350	222	Nancy + 1	France	100	France-Inter
		Madona + 1	USSR	50	
		Beograd-Studio B	Yugoslavia	10	
		Kuwait	Kuwait	200	
		Nouakchott	Mauretania	50	
		Sukhumi	USSR	30	
1359	221	Berlin	Germany (E)	250/100	Stimme Der DDR
		Tirane	Albania	120	
		Moskva	USSR	15	
		Bournemouth	UK	0.3	Local Radio
		Kirkuk	Iraq	240	
		Valencia	Spain		
1368	219	Yamit	Israel	20	
		Porto	Portugal	10	1st Programme
		Krakow + 1	Poland	60	2nd Programme
		Venezia + 13	Italy	25	3rd Programme
		Valjevo	Yugoslavia	10	
		Foxdale	UK	2	Manx Radio
1377	218	Lille 1	France	300	France Culture
		Lutsk + 1	Ukraine	50	

kHz	Metres	Station	Country	Power	Programme
1377	218	Kardzali	Bulgaria	30	1st Programme
contd		Kumrovec + 4	Yugoslavia	20	
		St. Pierre & Miquelon	St Pierre & Miquelon	4	
1386	216	Kaunas	USSR	1000	
		Athinai	Greece	50	2nd Programme
		Tuzla + 1	Yugoslavia	2	
		Orense ECS.18	Spain	2	RCE
		La Coruna EFJ.11	Spain	5	RCE
		Ahvaz	Iran	400	
1395	215	Lushnje	Albania	500	
		Malaga + 9	Spain	100	
		Hoogezand	Netherlands	10	Relay Lopik 3
		Leon EFE 5	Spain	2	RCE
		Alicante EFE.8	Spain	2	RCE
		Granada ECS.5	Spain	5	RCE
1404	214	Tripoli	Libya	20	
		Dniepro-petrovsk + 2	Ukraine	30	
		Bastia + 2	France	40	France Culture
		Baia Mare + 1	Roumania	15	
		Varazdin + 1	Yugoslavia	10	
		Helsinki 2	Finland	2	Swedish Programme
		Kipe	Guinea	400	
1413	212	Pristina	Yugoslavia	100	
		Bad Mergent-heim + 2	Germany (W)	3	Suddeutscher Rundfunk
		Sevilla ECS 8	Spain	5	RCE
		Barcelona EFJ 15	Spain	10	RCE
		Granada ECS 5	Spain	5	RCE
		Oviedo EFE 22	Spain	5	RCE
		Zaragoza EFJ 46	Spain	20	RCE
		Serona ECS 14	Spain	2	RCE
		Bejar EFJ 18	Spain		RCE
		Ile Masirah	Oman	1500	
1422	211	Saarbrucken	Germany (W)	1200/600	Saarlandischer Rundfunk
		Valmiera + 2	USSR	50	
		Alger 3	Algeria	40	
1431	210	Montemor-velho + 5	Portugal	100	
		Krivoi Rog	Ukraine	500	
		Skive	Denmark	70	3rd Programme
		Foggia + 2	Italy	2	2nd Programme
		Reading	UK	0.1	Radio 210
		Bernburg + 2	Germany (E)		Berliner Rundfunk
		Och	USSR	50	

kHz	Metres	Station	Country	Power	Programme
1440	208	Marnach	Luxembourg	1200	2nd Programme
		Kraljevo	Yugoslavia	100/10	
1449	207	Misurata	Libya	20	
		Coimbra	Portugal	1	1st Programme
		Kichinev + 1	USSR	50	
		Squinzano + 24	Italy	50	2nd Programme
		Karlovac + 1	Yugoslavia	10	
		Berlin	Germany (W)	5	Sender Freies Berlin
		Redmoss	UK	2	Radio 4 UK
1458	206	Lushnje	Albania	500	
		Torquay	UK	1	Radio 4 S West
		Birmingham	UK	10	Local Radio
		Carlisle	UK	0.5	Local Radio
		London	UK	50	Local Radio
		Manchester	UK	5	Local Radio
		Newcastle	UK	2	Local Radio
		Constanta 2	Roumania	50	
		Svetozarevo + 1	Yugoslavia	10	
1467	204.5	Monte-Carlo-Fontbonne	Monaco	400/200	
		Kiev + 1	Ukraine	300	
		Dededoruk	Turkey	10	
		Zvornik + 3	Yugoslavia	10	
		Isfahan	Iran	100	
		Frunze	USSR	30	
1476	203	Wien-Bisamberg	Austria	600	
		Lvov	Ukraine	120	
		Barcelona	Spain	25	
		Bilbao EFJ 43	Spain	10	RCE
1485	202	Bournemouth	UK	2	Radio 1
		Carlisle	UK	1	Radio 4 UK
		Brighton	UK	1	Local Radio
		Humberside	UK	2	Local Radio
		Merseyside	UK	2	Local Radio
		Oxford	UK	0.5	Local Radio
		Tours	France	1	France-Inter
		Aquila + 11	Italy	1	
		Bielsko Biala + 18	Poland	1	
		Anklam + 11	Germany (E)	1	DDR 1
		Saviese	Switzerland	1	
		Cesme + 4	Turkey	1	
		Orense EAJ 57	Spain		
		Dubai	United Arab Emirates		
1494	201	Rhodos	Greece	5	Regional Programme
		Ajaccio + 1	France	40	France Culture
		Edintsy + 1	USSR	25	
		Hulsberg	Netherlands	10	Relay Lopik 3
		Guarda	Portugal	0.02	

kHz	Metres	Station	Country	Power	Programme
1503	200	Stargard Szczecinski	Poland	300	2nd Programme
		Ulcinj + 1	Yugoslavia	10	
		Stoke-on-Trent	UK	0.5	Local Radio
		Beograd + 2	Yugoslavia	10	
		Burgos EFJ 52	Spain	3	RCE
		Pamplona EFE 57	Spain	2	RCE
		Bilbao	Spain		
		Nicosia	Cyprus	1	British Forces
		Alma-Ata	USSR	10	
1512	198	Chania	Greece	5	
		Beltem	Belgium	20	BRT 2 (Dutch)
		Pristina 2	Yugoslavia	100/20	
		Sotchi + 2	USSR	30	
		Aktiubinsk	USSR	100	
1521	197	Kosice + 6	Czechoslovakia	600	Hvezda & Regional
		Djedeida	Tunisia	100	
		Kalevala + 3	USSR	5	
		Nottingham	UK	0.5	Local Radio
		Oviedo EAJ.19	Spain	5	SER
		Pontevedra EAJ 40	Spain	3	
		Duba	Saudi Arabia	2000	
1530	196	Citta del Vaticano	Vatican	450	Radio Vatican
		Jitomar + 2	Ukraine	5	
		Funchal	Madeira	10	
1539	195	Voice of Peace		50	
		Mainflingen	Germany (W)	700	Deutschlandfunk
		Ialta + 2	Ukraine	25	
		Daugavpils + 2	USSR	5	
		Valladolid EAJ 47	Spain	5	
		Istok	Yugoslavia	2	
1548	194	Vinnitza	Ukraine	50	
		Edinburgh	UK	2	IBA R Forth
		Liverpool	UK	1	IBA R City
		London	UK	27.5	IBA Capital Radio
		Sheffield	UK	0.3	IBA R Hallam
		Bristol	UK	5	BBC Local Radio
		Cleveland	UK	1	BBC Local Radio
1557	193	Cyclops	Malta	600	
		Nice 2	France	300	France Culture
		Kaunas + 4	USSR	75	
		Osijek	Yugoslavia	50	
		Arbadil	Iran	10	
1566	191.5	Amalias	Greece	1.25	

kHz	Metres	Station	Country	Power	Programme
1566	191.5	Covilha	Portugal	1	1st Programme
contd		Sarnen	Switzerland	300	German Programme
		Leningrad	USSR	60	
		Odessa + 2	Ukraine	5	
		Smarje	Yugoslavia	2	
		Sfax	Tunisia	1200	National Programme (Arabic)
		Sandar Abbas	Iran	100	
1575	190.5	Madrid	Spain	5	RNE
		Genova + 17	Italy	50	3rd Programme
		Dresden + 3	Germany (E)		Berliner Rundfunk
		Porto	Portugal	10	3rd Programme
1584	189	Burgos + 28	Spain	1	
		Calvi + 1	France	1	France-Inter
		Branievo + 17	Poland	1	
		Bad Doberan + 8	Germany (E)	1	DDR 1
		Bar + 43	Yugoslavia	1	
		Leicester	UK	0.5	Local Radio
		Reus EAJ 11	Spain	2	SER
		Pamplona EAJ 6	Spain	2	SER
1593	188	Langenberg	Germany (W)	800	Westdeutscher Rundfunk
		Miscolc + 1	Hungary	20	
		Baneasa + 4	Roumania	15	
		Lisboa	Portugal	10	
1602	187	Sabadell EAJ 20	Spain	2	SER
		Bologna + 8	Italy	1	
		Cieszyn + 12	Poland	1	
		Angermuende + 12	Germany (E)	1	
		Alcira EAJ 54	Spain		SER

GEOGRAPHICAL LIST OF LONG AND MEDIUM
WAVE EUROPEAN STATIONS

AFGHANISTAN	kHz
Kabul	1278

ALBANIA	
Durres	1089
	1215
Korca	972
Kukes	999
Lushnje	1395
	1458
Shkoder	1314
Tirane	1359

ALGERIA	
Algiers	891
	981
	1422
Ain Beida	529
Bechar	576
Constantine	1305
Les Trembles	549
	1143
Tipaza	254

ANDORRA	
Radio Andorra	702
Sud Radio	819

AUSTRIA	
Graz-Dobl	1026
Innsbruck Aldrans	520
Klagenfurt	774
Wien-Bisamberg	585
	1476

AZORES	
Ponta Delgada	837

BELGIUM	
Kuurne	1188
La Louviere	1125
Liege	1233
Marche	1305
Veltem	1512
Waver-Overijse	540
	620
Wolvertem	927

BIELORUSSIA	kHz
Brest	1026
Minsk	281
	1197
Moghilev	1170
Mozyr	1008

BRAZIL	
Rio de Janeiro	936
	1224
	1296

BRITISH WEST INDIES	
Radio Paradise	1269

BULGARIA	
Biala Slatina	1161
Kardzali	1377
Petrich	747
Pleven	594
	945
Plovdiv	648
	864
Shumen	828
Sofia	774
	963
Stara Zagora	1125
Vidin	576
	1224

CANARY ISLANDS	
Las Palmas	837
	1008
	1215
Santa Cruz de Palma	1098
Santa Cruz Tenerife	621

CYPRUS	
Cape Greco	1233
Nicosia	603
	693
	1503
Paphos	918
Zakaki	639
	720
Zyyi	1323

CZECHOSLOVAKIA	kHz
Banska Bystrica	702
Bratislava	1098
Brno	900
	954
Brno Mesto	1332
Ceske Budejovice	846
Ceskoslovenko	272
	1287
Cheb	981
Jesenik	792
Kosice	1521
Nitra	1017
Mnich Hradiste	1071
Praha	639
	1233

DENMARK	
Kalundborg	245
	1062
Skive	1431
Torshavn	531

EGYPT	
Abis	774
Abu-Zabaal	558
	711
	873
Alexandria	1197
	1341
Assuit	1278
Asswan	1008
Batra	621
	819
	1107
Cairo	936
	1062
Tanta	1161

FINLAND	
Helsinki	558
	1404
Kuopio	756
Lahti	254
Mariehamn	603
Oulu	433
	540
Turku	963
	1278
Vaasa	1242

FRANCE	kHz
Ajaccio	1494
Allouis	164
Bastia	1404
Bordeaux	1206
Brest	1071
Calvi	1584
Lille	1377
Limoges	792
Lyons-Tramoyes	603
Marseille	675
	1242
Nancy	837
	1350
Nice	1557
Paris	585
	864
	963
Rennes	711
Strasbourg	1161
	1278
Toulouse	945
Tours	

GERMANY (EAST)	
Angermuende	1602
Anklam	1485
Bad Doberan	1584
Berlin	693
	1359
Bernburg	1431
Burg	263
	783
	1044
Dresden	1575
Erfurt	1170
Greifswald	531
Hoyerswerda	999
Koenigswusterhausen	603
	882
Leipzig	1323
Neubrandenburg	657
Oranienburg-Rehmate	182
Puttbus	729
Rostock	558
Schwerin	576

GERMANY (WEST)	
Aachen-Stolberg	702
Bad Mergentheim	1413
Berlin	567

Germany (West) — contd	kHz
Berlin (cont.)	810
	855
	1449
	990
Bodenseesender	666
Braunschweig	756
Bremen	936
Dannenberg	630
Donebach	155
Frankfurt	594
	873
Freiburg	828
Hamburg	972
Hannover	828
Heidelberg	711
Hof-Saale	520
	684
Holzkirchen	720
Langenberg	1593
Mainflingen	209
	1539
Munchen	1107
Munchen-Ismaning	801
	1197
Neumunster	1269
Nordkirchen	549
Saarbrucken	1422
Saarlouis-Felsburg	182
Stuttgart	576
	1143
Wolfsheim	1017

GREECE	
Amalias	1566
Athens	729
	1386
Chania	1512
Florina	1278
Ioannina	765
Iraklion	954
Kavalla	792
Kerkyra	1008
Larissa	945
Megara	981
Mesolongion	1071
Mytilini	1287
Orestias	1080
Rhodos	1260
	1494
Thessaloniki	1044
	1179

	kHz
Tripolis	1071
	1314
Zakynthos	927

GUINEA	
Kipe	1404

HUNGARY	
Budapest	873
Budapest-Lakihegy	1341
Miscolc	1593
Siofok	1251
Solt	540
Szolnok	1188

ICELAND	
Reykjavik	209

INDIA	
Calcutta	1134
Rajkot	1071

IRAN	
Abadan	1080
Ahvaz	1386
Arbadil	1314
	1557
Bonab	639
Farahabad	774
Isfahan	1467
Lar	1287
Mashad	684
Qashr' Shirin	612
Qeslagh	558
Rezaieh	936
Sandar Abbas	1566
Semnan	1008
	1296
Shiraz	981
	990
Tabriz	1026
	1152
Taybad	720
Teheran	1188
	1332
Yazd	1035

IRAQ	
Babylon	1035
Basra	693
Kirkuk	1359

	kHz		kHz
Iraq — contd		JORDAN	
Salman Pack	756	Amman	801
Thourah	909		855
IRELAND (EIRE)		KUWAIT	
Carraroe	540	Kuwait	1350
Cork	1188	Sulaibiyah	540
Dublin	1251		1134
Tullamore	567		
	612	LEBANON	
		Amchit	990
ISRAEL		Beyrouth	837
Beer-Sheva	1224		
Haifa	1206	LIBYA	
	1305	Benghazi	675
Jerusalem	531	El Beida	1125
	675	Misurata	1449
	711	Sebha	828
	1026	Tripoli	540
Tel Aviv	576		1053
	657		1251
	738		1404
	1242		
	1287	LUXEMBOURG	
Ufa	693	Junglinster	236
Yamit	1368	Marnach	1440
Zefat	846		
		MADEIRA	
ITALY		Funchal	1332
Ancona	1314		1530
Aquila	1485		
Bari	1116	MALTA	
Bologna	567	Cyclops	1557
	1098	Delimara	999
	1602		1341
Cagliari	1062	Radio Malta	756
Caltanissetta	191		
Foggia	1431	MARTINIQUE	
Genova	1575	Fort de France	1314
Messina	1143		
Milano	900	MAURETANIA	
	1035	Nouakchott	1350
Napoli	657		
Rimini	999	MONACO	
Roma	1332	Monte Carlo	218
San Remo	1188		702
Squinzano	1449		1467
Trieste	819		
	981	MOROCCO	
Venezia	1017	Agadir	936
	1368		1197
		Azilal	209

Morocco — contd	kHz		kHz
Casablanca	1080	Szczecin	1260
	1188	Warsaw	200
Ksar es Souk	864		227
Marrakech	972		819
	1152	Wroclaw	1206
Ouarzazate	1116		
Oujda	594	PORTUGAL	
	828	Braganca	576
Rabat	819	Castelo Branco	828
	1026	Chaves	1251
	1296	Coimbra	963
Safi	1026		1449
Sebaa Aioun	612	Covilha	1566
	702	Elvas	1332
	1044	Faro	558
Tanger	1233	Guarda	1494
Tantan	657	Lamego	927
Tetuan	1017	Lisboa	666
	1053		756
			1287
			1593
NETHERLANDS			
Amsterdam	1251	Miramar	783
Hengelo-Overijssel	891	Miranda do Douro	630
Hoogezand	1395	Mirandela	729
Hulsberg	1494	Montemorvelho	1431
Lopik	675	Norte	720
	747		1062
	1008	Portalegre	1197
		Porto	1170
			1368
NORWAY			1575
Bodoe	675		
Finmark	702	Porto Alto	1035
Oslo	218	Valenca do Minho	567
Stavanger	1314	Villa Real	621
Tromsoe	155	Viseu	693
Vigra	630		
		QATAR	
OMAN		Al Arish	954
Ile Masirah	1413		
Muscat	1242		
		ROUMANIA	
POLAND		Bacau	1179
Bielsko Biala	1485	Baia Mare	1404
Branievo	1584	Bailesti	720
Cieszyn	1602	Baneasa	1593
Katowice	1080	Brasov	155
Krakow	1368		567
Poznan	738	Bucuresti	855
Rzeszow	1305	Cluj	909
Stargard Szczecinski	1503		1152

Roumania — contd	kHz
Constantza	1314
	1458
Lasi	1053
Lugoj	756
Miercurea Ciuc	945
Oradea	603
Targa Jiu	558
Targa Mures	1323
Timisoara	630

ST PIERRE & MIQUELON

St Pierre & Miquelon	1377

SAUDI ARABIA

Damman	882
Duba	1521
Riyadh	585

SENEGAL

Dakar	765

SPAIN

Alcala Henares	1260
Alcira	1602
Alicante	1026
	1395
Badalona	1179
Barcelona	738
	828
	990
	1413
	1476
Bejar	1413
Bilbao	1134
	1503
	1476
Burgos	1503
	1584
Caceres	774
Cadiz	1314
Gijon	873
Granada	1080
	1395
	1413
La Coruna	639
	1386
Leon	1395
Madrid	585
	810
	918

Madrid — contd	kHz
	954
	1026
	1107
	1224
	1314
	1575
Malaga	1395
Mallorca	1080
Murcia	855
	1179
Orense	1386
	1485
Oviedo	729
	1413
	1521
Pamplona	1503
	1584
Pontevedra	1521
Reus	1584
Sabadell	1602
San Sebastian	1260
Serona	1413
Sevilla	684
	792
	1134
	1413
Valencia	1134
	1260
	1314
	1359
Valladolid	1107
	1539
Vigo	1026
Villareal	1224
Zamora	1134
Zaragoza	873
	1134
	1413

SUDAN

Sennar	1296
Soba	963

SWEDEN

Goeteborg	981
Hoerby	1179
Motala	191
Stockholm	774

SWITZERLAND

Beromunster	531

Switzerland — contd	kHz		kHz
Monte-Ceneri Passo	558	Odessa	765
Sarnen	1566		1278
Saviese	1485		1566
Sottens	765	Simferopol	648
		Tchernovtsy	657
SYRIA		Ujgorod	675
Homs	567		891
Damas-Sabbourah	666	Vinnitza	1548
Sarakeb	747		
Tartus	783	UNITED ARAB EMIRATES	
	1071	Abu Dhabi	810
		Dubai	1485
TUNISIA		Sadiyat	729
Djedeida	1521		
Gafsa	585	UNITED KINGDOM	
Sfax	720	Barnstaple	801
	1566		1053
Tunis	630	Barrow	693
	963		1053
		Belfast	1026
TURKEY		Bexhill	693
Ankara	182		1053
Antalya	891	Birmingham	1152
Cesme	1485		1452
Cukurova	630	Blackburn	855
Dedoruk	1467	Bournemouth	909
Diyabakir	1062		1359
Erzurum	245		1485
Etimesgut	200	Bradford	1278
Istanbul	1017	Brighton	693
Izmir	927		1053
Trabzon	954		1215
Umraniye	702		1485
		Bristol	1548
UKRAINE		Brookmans Park	909
Dniepropetrovsk	1071		1089
	1404		1215
Donetsk	711	Burghead	200
Ialta	1539		693
Jitomar	1530		810
Kharkov	837		1053
Kiev	209		1215
	783	Cambridge	1197
	1242	Carlisle	756
	1467		1458
Krivoi Rog	1431		1485
Lutsk	1377	Clevedon	909
Lvov	173	Cleveland	1548
	936	Crowborough	648
	1476	Derby	1116

United Kingdom — contd	kHz		kHz
Droitwich	200	Whitehaven	909
	693		1089
	1053	Wolverhampton	990
	1215	Wrexham	882
Dumfries	810		
Dundee	1053	USSR	
Edinburgh	1548	Achkabad	281
Exeter	693	Aktiubinsk	1512
	990	Alma Ata	182
Fareham	909		1098
	999		1341
	1089		1503
	1215	Astrakahn	792
Folkestone	693	Baku	218
	1053	Cheboksary	531
Foxdale	1368	Daugavpils	1539
Glasgow	1152	Djambul	1224
Gillingham	1035	Duchanbe	254
Hull	1053		702
	1215		1143
Humberside	1485		1161
Ipswich	1170		1296
Leeds	774	Edintsy	1494
Leicester	1584	Elista	846
Lisnagarvey	720	Engels	155
	909	Erevan	864
	1089	Frunze	1467
	1215	Gorkii	828
	1341	Ijevsk	594
Liverpool	1548	Iman	909
	1152	Iochkar-Ola	900
	1458	Irkutsk	263
	1548	Kalevala	1521
Stoke on Trent	1503	Kaliningrad	173
Swansea	1170		1116
Teeside	1170		1143
Torquay	909	Karaganda	747
	1458	Kaunas	1107
Tyneside	1152		1386
Tywyn	882		1557
	1089	Kazan	783
	1215	Kichinev	236
Washford	882		549
	1089		999
	1215		1449
Westerglen	200	Krasnoder	1089
	810	Kurgan	1053
	909	Leningrad	200
	1089		801
	1215		873

USSR — contd	kHz	YEMEN	kHz
Leningrad — contd	1125	Sanaa'a	1197
	1566		
Madona	1350	YUGOSLAVIA	
Medvejiegorsk	765	Aleksinac	1008
Mezen	918	Banja Luka	1071
Moscow	200	Bar	1584
	263	Beli Kriz	549
	846		1170
	1359	· Beograd	684
Murmansk	657		963
Nebit Dag	927		1350
Nikolaev	972		1503
Novosibirsk	272	Bor	981
Och	1431	Djakovica	936
Omsk	639	Gnjilane	1332
Orenburg	540	Istok	1539
Parnu	1332	Karlovac	1449
Perm	585	Koper-Belikriz	1080
Riga	576	Kraljevo	1440
	1071	Kumrovec	1377
Rostov	745	Loznica	1296
Saransk	1062	Ljubljana	918
Saratov	630	Maribor	558
Sotchi	1512	Murska Sobota	648
Stavropol	882	Nis	711
Sukhumi	1350	Nova Gradiska	1143
Tachkent	164	Novi Sad	837
Taldy Kurgan	245		1089
Tallinn	612		1107
	711		1269
	1035	Osijek	1557
Tartu	1215	Pristina	1512
Tbilissi	191		1413
	1044	Sarajevo	612
Tcheliabinsk	738		747
	855	Skopje	810
Tiumen	603		1314
Tselinograd	684		1566
	999	Smarje	1566
Ukhta	621	Sombor	666
Ulan-Ude	801	Split	774
Valmiera	1422	Svetozarevo	1458
Vilnius	666	Titograd	882
Volgograd	567	Tuzla	1386
	810	Ulcinj	1503
	945	Valjevo	1368
Voronej	775	Varazdin	1404
		Zagreb	1062
			1125
VATICAN			1134
Vatican City	1530	Zvornik	1467

In general, short-wave stations adjust their frequency schedules four times a year, because of different propagation conditions in spring, summer, autumn and winter. Alterations are arranged on an international basis.

Although some stations may use virtually the same channels throughout the year with only minor differences, others use particular frequencies during only one or two of the four periods. The short-wave list therefore has columns marked M, M, S and N which indicate the four plan periods used by the ITU for the HF Tentative Schedules commencing March, May, September and November respectively when channelling is changed. The underline symbol (_) indicates that the station was planned for the period in the previous year's tentative schedule, and the oblique stroke (/) shows that the station was included in the plans for the year starting in March 1979. An X in these columns shows that a station originally allocated for that plan period has, during the period, been deleted from the plan. The list also includes stations, not shown in the Tentative Schedule, which are known to be active or which are expected to operate in the near future. Some frequencies listed may only be audible in favourable reception conditions, and others may only be in use seasonally. Stations that are known to transmit broadcast programmes, using ISB or SSB transmission modes, which are not intended for general reception by the public, are identified by the addition of an asterisk * after the station name or if the site is unknown after the country code. Other changes noted by the listener can be recorded similarly. The columns also indicate the extent of each short-wave band allocated to broadcasting: these indications exclude the out-of-band frequencies which are also occasionally used.

Transmitter power in the short-wave bands is not easily defined, because the majority of stations have a number of senders of varying power, any one of which may be used as required. The powers quoted are therefore the lowest and highest known to operate at a location over 24 hours and should be used only as a rough guide, because it is impossible to cover all the possibilities.

A high-gain aerial, beamed towards the listener, can provide a strong signal from a comparatively low-powered transmitter, although a narrow-beam array, powered with 250 kW but directed away from the receiving site, may be barely audible. Thus power figures merely indicate the *capability* of a station in terms of field strength: the direction of main radiation may or may not favour a listener outside the target zone.

A station name can be that of the large town nearest to the transmitting site, or it can be the capital of the country even, although there may be more than one transmitting site. Occasionally two different place names are given, separated by an oblique stroke; this indicates that the channel is shared. Where the same transmitter operates at different times on adjacent channels, separate entries are made; this accounts for the multiplicity of entries under some place names. In

most cases clandestine stations are listed by their slogan only and no country is shown.

This list of stations is compiled from information obtained from broadcasting authorities and the BBC Receiving Station, Caversham Park, Reading covering the period November 1978—March 1979.

A geographical list of short-wave stations will be found on page 179.

MHz	Metres	Station	Country	kW
2.500	120.00	Rugby Std Freq	UK	
2.670	112.36	Sariwon	Korea (N)	
2.739	109.53	Riyadh*	Saudi Arabia	
2.745	109.29	Sinuiju	Korea (N)	
2.765	108.50	Pyongyang	Korea (N)	
2.775	108.11	Hamhung	Korea (N)	
2.850	105.26	Pyongyang	Korea (N)	
3.000	100.00	Fukien Front Stn	China Rep	
3.015	99.50	Pyongyang	Korea (N)	120
3.030	99.01	Wonsan	Korea (N)	
3.155	95.09	Peshawar	Pakistan	10
3.195	93.90	Baghdad	Iraq	50
3.200	93.75	Fukien Front Stn	China Rep	
3.205	93.60	Ibadan	Nigeria	10
		Lucknow	India	10
3.210	93.46	Maputo	Mozambique	
3.215	93.31	Rawalpindi	Pakistan	10
3.220	93.17	Peking	China Rep	120
3.222	93.11	Lama Karma	Togo	10
3.223	93.08	Simla	India	2.5
		Mbabne	Swaziland	
3.225	93.02	Tovar	Venezuela	1
		Lins	Brasil	0.5
3.227	92.97	Monrovia	Liberia	10
3.230	92.88	Johannesburg	South Africa	100/250
3.232	92.82	Brazzaville	Congo	25
		Tananarive	Malagasy Rep	30
3.235	92.74	Gauhati	Gauhati	10
		Marilia	Brasil	0.5
3.240	92.59	Islamabad	Pakistan	10/100
		Lima	Peru	
3.242	92.54	Baghdad	Iraq	50
3.245	92.45	Caracas	Venezuela	1
3.250	92.31	Bloemendal	South Africa	20

MHz	Metres	Station	Country	kW
		Tovar	Venezuela	
		El Tigre	Venezuela	1
3.255	92.17	Monrovia	Liberia	10
3.260	92.02	Niamey	Niger	4
		Kweiyang	China Rep	
3.265	91.88	Georgetown	Guyana	2
		L Marques	Mozambique	25/100
			Brasil	1
3.270	91.74	Peking	China Rep	
3.273	91.66	Quetta	Pakistan	
3.277	91.55		India	
3.285	91.32	Manila	Philippines	2.5
		Meyerton	South Africa	
		Pernambuco	Brasil	1
		Puerto Cabello	Venezuela	1
3.288	91.24	Tananarive	Malagasy Rep	100
3.290	91.19	Tristan da Cunha	Tristan da Cunha	0.04
		Peking	China Rep	
3.295	91.05	Delhi	India	20
		Trujillo	Venezuela	
		Lusaka	Zambia	10
		Accra	Ghana	10
3.300	90.91	Libreville	Gabon Rep	4/20
		Bujumbara	Burundi	25
		Guatemala	Guatemala	10
		Belmopan	Honduras Br	1
3.305	90.77	Daru	Papua	
3.307	90.72	Gwelo	Rhodesia	10/100
3.315	90.50	Bhopal	India	10
		Fort de France	Martinique	4
3.316	90.47	Freetown	Sierra Leone	10
3.320	90.36	Bloemfontein	South Africa	
		Pyongyang	Korea (N)	100
3.325	90.23	Maceio	Brasil	25
3.330	90.09	Dzaudazi	Comoro Is	15
		Peshawar	Pakistan	10
		Kigali	Rwanda	5
3.335	89.96	Wewak	Papua New Guinea	10
3.336	89.93	Ziguinchor	Senegal	4
3.338	89.87	Maputo	Mozambique	
3.339	89.85	Zanzibar	Tanzania	10
3.340	89.82	Kampala	Uganda	7.5
3.343	89.74	Nampula	Mozambique	
3.345	89.69	Manila	Philippines	40
		Uberlandia	Brasil	5
		Luanda	Angola	
3.346	89.66	Lusaka	Zambia	120
3.350	89.55	Tema	Ghana	20

MHz	Metres	Station	Country	kW
		Franceville	Gabon Rep	20
3.355	89.42	Noumea	New Caledonia	20
		Gaberone	Botswana	10
		Valencia	Venezuela	1
		Kurseong	India	20
		Luanda	Angola	
3.360	89.29	Milne Bay	Papua New Guinea	10
		Peking	China Rep	
3.365	89.15	Delhi	India	10
		Araraquara	Brasil	1
3.366	89.13	Tema	Ghana	10
3.370	89.02	Tananarive	Malagasy Rep	4
		Beira	Mozambique	10
3.375	88.89	Luanda	Angola	10
		Gauhati	India	10
3.380	88.76	Blantyre	Malawi	100
		Esmeraldas	Ecuador	10
3.385	88.63	Cayenne	Guyana Fr	4
		Colombo	Ceylon	10
		Rabaul	Papua New Guinea	10
		Barcelona	Venezuela	1
3.388	88.55	Bloemendal	South Africa	
3.390	88.50	Kabul	Afghanistan	100
		Peking	China Rep	20/120
		S Domingo	Ecuador	5
3.395	88.37	Colombo	Ceylon	
		Merida	Venezuela	1
3.396	88.34	Kaduna	Niger	10
		Gwelo	Rhodesia	100
3.397	88.31		Pakistan	
3.400	88.24	Fukien Front Stn	China Rep	
3.417	87.80	Met Station	USSR	
3.421	87.69	Medan	Indonesia	1
3.425	87.59	Khumaltar	Nepal	100
3.440	87.21	Met Station*	USSR	
3.450	86.96	Peking	China Rep	
3.460	86.71		USSR	
3.470	86.46	Met Station*	USSR	
3.500	85.71	Peking	China Rep	20/240
3.535	84.87	Fukien Front Stn	China Rep	
3.550	84.51	R Dili	Timor	
3.560	84.27	Pyongyang	Korea (N)	120
3.640	82.42	Fukien Front Stn	China Rep	
3.660	81.97	Peking	China Rep	
3.695	81.19	Pyongyang	Korea (N)	
3.700	81.08	Peking	China Rep	
3.830	78.33	Peking	China Rep	120

MHz	Metres	Station	Country	kW
3.850	77.92	Dili	Timor	
3.860	77.72	Karachi	Pakistan	
3.885	77.22	C Verde Is	Cape Verde Is	
3.890	77.12	Pyongyang	Korea (N)	
		Karachi	Pakistan	10
3.900	76.92	Hailar	Mongolian Rep	
		Ventiane	Laos	
3.905	76.82	Delhi	India	20
		Palang	Indonesia	
3.910	76.73	Tokio	Japan	10
		Surakarta	Indonesia	
3.915	76.63	Islamabad	Pakistan	10
		Tebrau	Malaysia	100/250
		Kashmir Radio		
3.920	76.53	Peking	China Rep	
3.925	76.43	Delhi	India	10
		Port Moresby	Papua New Guinea	2
		Tokio	Japan	50
		Jakarta	China Rep	
3.930	76.34	Huhetot	Mongolian Rep	
		Barlavento	Cape Verde Is	10
		Met Station*	USSR	
		Tokio	Japan	
			Korea (S)	
3.935	76.24	Semarang	Indonesia	10
3.940	76.14	Wunan	China Rep	
			Indonesia	
			USSR	
3.945	76.05	Denpassar	Indonesia	10
		Hokkaido	Japan	
3.950	75.95	Peking	China Rep	
		Baghdad	Iraq	100
		Jermate	Indonesia	
3.952	75.91	London	UK	100
		Peking	China Rep	
3.955	75.85	Bloemendal	South Africa	20
		Padang	Indonesia	
		Warszawa	Poland	
3.960	75.76	Peking	China Rep	120
		Padang	Indonesia	10
		Holzkirchen	Germany (W)	10
		Biblis	Germany (W)	100
		Lampertheim	Germany (W)	
		Vladivostock	USSR	
3.965	75.66	Bloemendal	South Africa	20
		Pontianak	Indonesia	10
		Allouis	France	4
		Karachi	Pakistan	

MHz	Metres	Station	Country	kW
3.970	75.57	London	UK	
		Buea	Cameroon	8
		Huhetot	Mongolian Rep	
		Holzkirchen	Germany (W)	10
		Riyadh	Saudi Arabia	
3.975	75.47	London	UK	
3.980	75.38	Ismaning	Germany (W)	
		Bloemendal	South Africa	100
		Surabaya	Indonesia	
3.985	75.28	Schwarzenburg	Switzerland	250
		Riobamba	Ecuador	1
		Peking	China Rep	
		Lampertheim	Germany (W)	20
		Biblis	Germany (W)	
3.990	75.19	Limassol	Cyprus	
		Monrovia	Liberia	250
		Biblis	Germany (W)	100
		Lampertheim	Germany (W)	20
		Holzkirchen	Germany (W)	10
			USSR	
		Urumchi	China Rep	
3.995	75.09	Vladivostock	USSR	
		Roma	Italy	5
		Julich	Germany (W)	100
		Bloemendal	South Africa	20
		Pontianak	Indonesia	
3.999	75.02	Godthaab	Greenland	1
4.000	75.00	Kabul	Afghanistan	100
		Hanoi	Vietnam	
		Phnom Penh	Cambodia	
4.005	74.91	Pontianak	Indonesia	
4.010	74.81	Frunze	USSR	15
4.020	74.63	Peking	China Rep	120
		Islamabad	Pakistan	10/100
4.030	74.44	Magadan	USSR	
		Kanggye	Korea (N)	
4.035	74.35	Lhasa	Tibet	
4.040	74.26	Erevan	USSR	
		Vladivostok	USSR	
4.045	74.17	Fukien Front Stn	China Rep	
4.050	74.07	Frunze	USSR	
		Yuzhno Sakhalinsk	USSR	
4.055	73.98	Petropavlovsk	USSR	50
4.060	73.89	Rawalpindi	Pakistan	
4.068	73.75	Huhetot	Mongolian Rep	
			China Rep	
4.070	73.71	Jakarta	Indonesia	
4.080	73.53	Ulan Bator	Mongolian Rep	50
4.110	72.99	Urumchi	China Rep	

MHz	Metres	Station	Country	kW
4.115	72.90	V Rev Pty for Reunification		
4.130	72.64	Peking	China Rep	
4.180	71.77	Peking	China Rep	
4.190	71.60	Urumchi	China Rep	
4.200	71.43	Peking	China Rep	120
4.220	71.09	Urumchi	China Rep	
4.250	70.59	Peking	China Rep	120
4.284	70.03	Vinh Phu	Vietnam	
4.320	69.44	Vientiane	Laos	
4.330	69.28	Fukien Front Stn	China Rep	
4.380	68.49	Fukien Front Stn	China Rep	
4.395	68.26	Yakutsk	USSR	50
4.410	68.03		China Rep	
4.425	67.80		USSR	
4.460	67.26	Peking	China Rep	
4.465	67.19	Vladivostock	USSR	
4.467	67.16	London	UK	
4.475	67.04	RFE/R Liberty*		
4.485	66.89	Petropavlovsk	USSR	
		Ufa	USSR	
4.500	66.67	Urumchi	China Rep	
4.505	66.59	RFE/R Liberty*		
4.520	66.37	Palana	USSR	
		Khanty-Mansiysk	USSR	
4.525	66.30	Ikechao/Tungsheng	Mongolian Rep	
		Standard Time Signal*	Germany (W)	
4.545	66.01	Alma Ata	USSR	50
4.557	65.83	V Rev Pty for Reunification		
4.565	65.72	RFE/R Liberty*		
4.600	65.22	Thu Dau Mot City	Vietnam	
4.603	65.17	Xieng Khouang	Laos	
4.620	64.94	Peking	China Rep	
4.630	64.79	Peking	China Rep	
4.635	64.72	Duchanbe	USSR	50
4.645	64.59	Vientiane	Laos	
			USSR	
		Met Station*	USSR	
4.647	64.56	Ha Bac	Vietnam	
4.650	64.52	Houa Phan	Laos	
4.654	64.46		USSR	
4.656	64.43	Met Station	USSR	
4.665	64.31	Houa Phan	Laos	
		Met Station	USSR	
4.675	64.17	V of NUFK		
4.678	64.13	Met Station*	USSR	
4.679	64.12	Espejo	Ecuador	

MHz	Metres	Station	Country	kW
4.680	64.10	Hue	Vietnam	
4.685	64.03	Met Station*	USSR	
4.692	63.94	Quang Ninh	Vietnam	
4.695	63.90	Met Station	USSR	
		RFE/R Liberty*		
4.700	63.83	Luang Prabang	Laos	
		Surabaya	Indonesia	
4.710	63.69		USSR	
4.720	63.56		USSR	
		Jakarta	Indonesia	
		Karachi	Pakistan	
		Bassacongo	Angola	0.5
		Sao Vicente	Cape Verde Is	1.5
4.725	63.49	Rangoon	Burma	50
4.730	63.42		USSR	
4.735	63.36	Karachi	Pakistan	
4.747	63.20		USSR	
4.750	63.16	Lubumbashi	Zaire	10
		Bertoua	Cameroon	20
4.755	63.09	Bogota	Colombia	11
		Jakarta	Indonesia	
4.760	63.03	Peking	China Rep	
		Mbabane	Swaziland	
		Dzhambul	USSR	
4.762	63.00	Gia Lai-Cong Tum	Vietnam	
4.763	62.99	Ulan Bator	Mongolian Rep	50
4.764	62.97	Medan	Indonesia	
4.765	62.96	Guayaquil	Ecuador	5
		Brazzaville	Congo	50
			USSR	
4.770	62.89	Peking	China Rep	
		Monrovia	Liberia	10
		Son La	Vietnam	
		Bolivar	Venezuela	1
		Pyongyang	Korea (N)	
		Jakarta	Indonesia	
4.775	62.83	Gauhati	India	10
		Kabul	Afghanistan	100
		Sibolga	India	50
		Libreville	Gabon Rep	100
		Cuiba	Brasil	
4.780	62.76	Djibuti	Afars & Issas	4
		Petrozavodsk	USSR	50
		Valencia	Venezuela	
		Luambo	Angola	
4.783	62.72	Bamako	Malawi	18
4.785	62.70	Baku	USSR	50
		Ibaque	Colombia	
		Dar-es-Salaam	Tanzania	50

MHz	Metres	Station	Country	kW
		Sao Luiz	Brasil	
		Yunnan	China Rep	
		Cao Lang	Vietnam	
4.790	62.63	Iquitos	Peru	
		Caracas	Ecuador	
		Ulan Bator	Mongolian Rep	
			Swaziland	
			Angola	
4.795	62.57	Ulan Ude	USSR	50
		Aquidauana	Brasil	
		R Neuva America	Bolivia	1
		Brazzaville	Congo	4
4.800	62.50	Peking	China Rep	
		Hyderabad	India	10
		Barquesimeto	Venezuela	10
		Maseru	Lesotho	10
4.804	62.45	Nairobi	Kenya	5
4.805	62.43	Djakarta	Indonesia	100
		Manaus	Brasil	10
		Yacuiba	Bolivia	
4.807	62.41	Sao Thome	St Thomas Is	10
		Santiago	Dominican Rep	
4.810	62.37	Erevan	USSR	
		Maracaibo	Venezuela	2
		Bloemfontein	South Africa	
4.815	68.31	Peking	China Rep	
		Londrina	Brasil	
		Ouagadougou	Upper Volta	20
		Vallendupar	Colombia	
			USSR	
4.820	62.24	Luanda	Angola	100
			USSR	50
		Barquesimeto	Venezuela	1
		Ha Tuyen	Vietnam	
		Tegucigalpa	Honduras Rep	1
		Calcutta	India	10
4.825	62.18	Moskva	USSR	100
		Achkhabad	USSR	5
		Dar-es-Salaam	Tanzania	
		Bamako	Malawi	18
		Braganca	Brasil	5
4.828	62.14	Gwelo	Rhodesia	100
4.830	62.11	Franceville	Gabon Rep	20
			USSR	
		Ulan Bator	Mongolian Rep	
4.832	62.09	San Jose	Costa Rica	1
		Bangkok	Thailand	10
4.835	62.05	Kuching	Malaysia	10
		Meyerton	South Africa	

MHz	Metres	Station	Country	kW
		Boa Vista	Brasil	
			USSR	
4.840	61.98	Bombay	India	10
		Valera	Venezuela	1
		Heeilunkiang	China Rep	
			USSR	
4.843	61.95	Point Noire	Congo	4
4.845	61.92	Gaberones	Botswana	10
		La Paz	Bolivia	
		Bucramanga	Colombia	
		Nouakchott	Mauretania	100
		Kuala Lumpur	Malaysia	50
4.850	61.86	Moskva	USSR	
		S Domingo	Dominican Rep.	3
		Tashkent	USSR	
		P Louis	Mauritius	10
		Yaounde	Cameroon	
		Paramaribo	Surinam	
		Peking	China Rep	
			India	
		Ulan Bator	Mongolian Rep	
4.853	61.82	Sanaa	Yemen	
4.855	61.79	Palembang	Indonesia	10
		Taubate	Brasil	
		L Marques	Mozambique	20
4.860	61.73	Delhi	India	10
		Moskva	USSR	
		Tchita	USSR	
		Saurimo	Angola	5
		Maracaibo	Venezuela	10
		Hanoi	Vietnam	
4.865	61.66	Belem	Brasil	
		Arauca	Colombia	1
		Lanchow	China Rep	
		Maputo	Mozambique	7.5
		Berakas	Brunei	10
4.870	61.60	Caracas	Venezuela	
		Cotonou	Benin	30/50
		Ekala	Ceylon	10
4.872	61.58	Peking	China Rep	
4.875	61.54	Bloemfontein	South Africa	100/250
		Rio de Janeiro	Brasil	
		Bamako	Malawi	
		S Crux del Sur	Bolivia	10
		Medellin	Colombia	
		Ibaque	Colombia	5
		Jakarta	Indonesia	
4.880	61.48	Barquisemeto	Venezuela	10
		Peking	China Rep	

MHz	Metres	Station	Country	kW
		Bloemendal	South Africa	
		V of Kawthulay		
		Dacca	Bangladesh	
4.881	61.46	Thanh Hoa	Vietnam	
4.882	61.45		Germany (W)*	
4.885	61.41	Pocas de Caldas	Brasil	
		Villavicencio	Colombia	1
		Peking	China Rep	
		Mombasa	Kenya	10
		Bukittinggi	Indonesia	
4.890	61.35	Caracas	Venezuela	
		Port Moresby	Papua New Guinea	10
		Dakar	Senegal	25
		Dacca	Bangladesh	
4.895	61.29	Tyumen	USSR	50
		Achkhabad	USSR	50
		Kuching	Malaysia	10
		Beira	Mozambique	100
		Kurseong	India	
		Huhetot	Mongolian Rep	
		Peking	China Rep	
4.896	61.27	Silva Porto	Angola	1
4.900	61.22	Barquisemeto	Venezuela	10
		Cordac	Burundi	2.5
		Ekala	Ceylon	
4.905	61.16	Peking	China Rep	
		Fort Lamy	Chad	100
		Rio de Janeiro	Brasil	
		Gedja	Ethiopia	100
4.908	61.12	Phnom Penh	Cambodia	
		V of Kampuchea People		
4.910	61.10	Conakry	Guinea	18
		Carora	Venezuela	
		Quito	Ecuador	10
		Lusaka	Zambia	10
4.915	61.04	Macapa	Brasil	
		Accra	Ghana	
		Langata	Kenya	100
		Nanning	China Rep	
		Gutapuri/ Valledupar	Colombia	
4.920	60.98	Brisbane	Australia	
		Caracas	Venezuela	
		Madras	India	
		Kiev	Ukraine	
		El Progreso	Honduras Rep	10
		Quito	Ecuador	5

MHz	Metres	Station	Country	kW
		Jakarta	Indonesia	
4.923	60.94	Quito	Ecuador	
4.925	60.91	Maputo	Mozambique	7.5
		Bata	Guinea	5
		Yaounde	Cameroon	
			Brasil	
		Aranca	Colombia	
4.927	60.89	Djambi	Indonesia	7.5
4.930	60.85		USSR	
		San Cristobal	Venezuela	
		Quito	Ecuador	
		Jakarta	Indonesia	
4.932	60.83	Jakarta	Indonesia	10
		Benin	Niger	10
		Hanoi	Vietnam	
4.934	60.80	Nairobi	Kenya	5
4.935	60.79	Tarapoto	Peru	
		Rawalpindi	Pakistan	10
4.940	60.73	Abidjan	Ivory Coast	25
		Kiev	Ukraine	50
		San Filipe	Venezuela	10
		Colombo	Ceylon	10
		Peking	China Rep	
		Hanoi	Vietnam	
4.945	60.67	Neiva	Colombia	2.5
		Pocas de Caldas	Brasil	5
		Hanoi	Vietnam	
		Ekala	Ceylon	
4.950	60.61	Nairobi	Kenya	10
4.952	60.58	Silinhot	Mongolian Rep	50
4.955	60.54	Bogota	Colombia	50
			USSR	
		Kuching	Malawi	
4.957	60.52	Baku	USSR	50
4.960	60.48	Cumana	Venezuela	1
		Jakarta	Indonesia	
		Peking	China Rep	
		Sucua	Ecuador	
		Delhi	India	
			Venezuela	
4.965	60.42	Bogota	Colombia	5
		Potosi	Bolivia	
		Uberaba	Brasil	
		Lusaka	Zambia	2.5
4.970	60.36	Villa de Cura	Venezuela	10
		Urumchi	China Rep	
		Jurong	Senegal	10
4.972	60.34	Yaounde	Cameroon	30
		Koya Kinabalu	Malaysia	10

MHz	Metres	Station	Country	kW
		Cayenne	Guyana Fr.	1
4.975	60.30	Blagoveshchensk	USSR	50
		Foochow	China Rep	
		Lima	Peru	
		Sabah	Malaysia	
		Sanaa	Yemen	
4.976	69.29	Kampala	Uganda	7.5
		Sao Luis	Brasil	
4.980	60.24	San Cristobal	Venezuela	10
		Tema	Ghana	
			Burma	
		Peking	China Rep	
4.985	60.18	Luanda	Angola	0.5
		Tananarive	Malagasy Rep	4
		Goiana	Brasil	5
		Penang	Malaysia	10
4.990	60.12	Alma Ata	USSR	50
		Erevan	USSR	
		Barquesimeto	Venezuela	15
		Lagos	Niger	50
		Choraya	Bolivia	
		Meyerton	South Africa	250
		Changsha	China Rep	
4.995	60.06	Hanoi	Vietnam	
		Goiana	Brasil	
		Andina	Peru	
		Ulan Bator	Mongolian Rep	
4.999	60.01	Bangui	Central African Rep	100
5.000	60.00	Rugby Std Freq	UK	
		Boulder " "	USA	
		Honolulu " "	Hawaii	
		Turin	Italy	
		Rome	Italy	
5.005	59.94	La Paz	Bolivia	
		Lalitpur	Nepal	100
5.010	59.88	Santo Domingo	Dominican Rep	
		Garoura	Cameroon	30
		Jurong	Senegal	10
		Islamabad	Pakistan	10/100
		Tananarive	Malagasy Rep	4
5.015	59.82	Arkhangelsk	USSR	15
		Vladivostock	USSR	50
		Rio de Janeiro	Brasil	
5.016	59.81	Gwelo	Rhodesia	100
5.020	59.76	Niamey	Niger	20
		Caracas	Venezuela	
5.025	59.70	Aquidauana	Brasil	
		Andahuaylas	Peru	

MHz	Metres	Station	Country	kW
5.026	59.69	Kampala	Uganda	7.5
5.030	59.64	Kuching	Malaysia	10
		Caracas	Venezuela	10
		Urumchi	China Rep	
5.035	59.58	Sao Paulo	Brasil	
		Tachkent	USSR	
		Florencia	Colombia	
5.038	59.55	Khartoum	Sudan	20
5.040	59.52	Villa Vicencio	Colombia	3
		Tbilisi	USSR	50
		Bissau	Guinea	
5.045	59.46	Rarotonga	Cook Is	1
		Prudente	Brasil	
5.047	59.44	Lome	Togo	100
5.050	59.41	Dar-es-Salaam	Tanzania	20
		Caracas	Venezuela	10
		Jakarta	Indonesia	
5.052	59.38	Jurong	Senegal	
5.054	59.36	San Jose	Costa Rica	
5.055	59.35	Ulan Bator	Mongolian Rep	50
		Cuiba	Brasil	
5.057	59.32	Tirane	Albania	
5.060	59.29	Aden	Yemen	7.5
		Urumchi	China Rep	
5.061	59.28	Islamabad	Pakistan	
		Quito	Ecuador	
5.065	59.23	Petrozavodsk	USSR	
5.075	59.11	Bogota	Colombia	25
		Peking	China Rep	
5.090	58.94	Peking	China Rep	120
5.095	58.88	Sutatenza	Colombia	50
5.110	58.71	V of People of Burma		
5.112	58.69		Pakistan	
5.125	58.54	RFE/R Liberty*		
		Peking	China Rep	
5.135	58.42	Peking	China Rep	
5.139	58.38	Phu Khanh	Vietnam	
5.145	58.31	Peking	China Rep	
5.160	58.14	Vientiane	Laos	
5.163	58.10	Peking	China Rep	
5.189	57.81	Rangoon	Burma	
5.195	57.75	Julich*	Germany (West)	
5.220	57.47	Peking	China Rep	
5.230	57.36		USA*	
5.240	57.25	Fukien Front Stn	China Rep	
5.250	57.14	Peking*	China Rep	
5.255	57.09		USSR*	
5.260	57.03	Alma Ata	USSR	

MHz	Metres	Station	Country	kW
		Riyadh*	Saudi Arabia	
5.265	56.98	Fukien Front Stn	China Rep	
5.290	56.71	Krasnoiarsk	USSR	
			USSR*	
5.295	56.66	Peking	China Rep	
		RFE/R Liberty*		
5.320	56.39	Peking	China Rep	
			USSR	
5.339	56.19	Sao Tome	St Thomas Is	
5.345	56.13	R Freedom from S Yemen		
5.390	55.66	Riyadh*	Saudi Arabia	
5.420	55.35	Peking	China Rep	
5.440	55.15	Urumchi	China Rep	
5.455	55.00		USSR*	
5.460	54.95	Tangier*	Morocco	
5.470	54.84		USSR*	
5.504	53.63	Hoang Lien Son	Vietnam	
5.610	53.48	Luanda	Angola	
5.700	52.63		USSR	
5.703	52.60		USA*	
5.745	52.22	Greenville*	USA	
5.770	51.99	Moscow	USSR	
5.790	51.81	RFE/R Liberty*		
5.794	51.78		USSR	
5.815	51.59		USSR*	
5.830	51.46		USSR*	
			Germany (W)*	
5.840	51.37	Peking	China Rep	
5.845	51.33	RFE/R Liberty*		
5.850	51 28	Peking	China Rep	
5.860	51.19	Peking	China Rep	
5.870	51.11	Pyongyang	Korea (N)	
5.872	51.09	London*	UK	
5.876	51.06	Riyadh	Saudi Arabia	
5.880	51.02	Peking	China Rep	
		Jakarta	Indonesia	
5.882	51.00	Jerusalem	Israel	
		Buenos Aires	Argentina	
5.885	50.98	Jakarta	Indonesia	
			USSR	
5.900	50.85	Fukien Front Stn	China Rep	
		Jerusalem	Israel	
		Moskva	USSR	
5.905	50.80	Moskva	USSR	
		Tula	USSR	
5.910	50.71	Moskva*	USSR	50
		Vologda	USSR	
5.914	50.73	Bizam Radio		

MHz	Metres	Station	Country	kW	M	M	S	N
5.915	50.72	Armavir	USSR					
		Kiev	Ukraine					
		Jerusalem	Israel					
		Peking	China Rep					
		Sofia	Bulgaria					
		Alma Ata	USSR					
5.920	50.68		USSR	50				
		Sverdlovsk	USSR	100				
5.925	50.63	Tashkent	USSR	50				
		Wien	Austria					
		Urumchi	China Rep					
5.927	50.61	Lai Chau	Vietnam					
5.930	50.59	Murmansk	USSR	15				
		Tbilisi	USSR					
		Prague	Czechoslovakia	100				
5.935	50.55	Riga	USSR	50				
			USSR					
		Peking	China Rep					
		Lhasa	Tibet					
5.940	50.51	Magadan	USSR					
		Moskva	USSR					
		Petropavlovsk	USSR					
		Sofia	Bulgaria					
5.945	50.46	Minsk	USSR					
		Tashkent	USSR					
		Monte Carlo	Monaco					
		RFE/R Liberty*						
		Wien	Austria					
		Tirane	Albania					
5.950	50.42	Peking	China Rep					
			USSR					
		Leningrad	USSR					
		Managua	Nicaragua					
		V of Iraqi Kurdistan						
5.955	50.38	London	UK	100	_			
		Dixon	USA	100/250	∠	_	_	_
		Bluefields	Nicaragua	0.5	∠	_	_	_
		Llallagua	Bolivia	1	∠	_	_	_
		Pakanbaru	Indonesia	10	∠	_	_	_
		Allouis	France	100	∠	_	_	_
		Pt Limon	Costa Rica	1	∠	_	_	_
		S Paulo	Brasil	7.5	∠	_	_	_
		Lopik	Netherlands	100	∠	_	_	_
		Tinang	Philippines	250		_	_	_
		Kavalla	Greece	250	∠	_	_	_
		Athinai	Greece	100	/			
		Nauen	Germany (E)	50/500	∠			_
		Manzini	Swaziland	25	_			

MHz	Metres	Station	Country	kW	M	M	S	N
		Komsomolskamur	USSR	100				
		Diosd	Hungary	100	_			
		Santiago	Chile	1	/		_	_
		Ismaning	Germany (W)	100	/	_	_	
		Biblis	Germany (W)	100		_		_
		Holzkirchen	Germany (W)	10	/	_		
		Lampertheim	Germany (W)	100	L	_	_	_
		Villavicencio	Colombia	5	L	_	_	_
		Shepparton	Australia	100			_	_
		Franceville	Gabon	500			_	
5.960	50.34	Antigua	Br W Indies	250			_	_
		Jammu	India	1	L	_	_	_
		Delhi	India	100	_	_	_	_
		Godthaab	Greenland	1/10	L	_	_	_
		Sisoguichi	Mexico	0.3	L	_	_	_
		Alma Ata	USSR	100	L	_	_	_
		Armavir	USSR	100		_	_	
		Sverdlovsk	USSR	100	L		_	
		Blagoevechtchen	USSR	100		_	_	
		Vladivostock	USSR	100	L	_	_	_
		Sackville	Canada	250	L	_	_	_
		Tirane	Albania					
		S Rosa Copan	Honduras Rep	1	/		_	_
		Warszawa	Poland	1	_			
		Wien	Austria	100	_	_	_	_
		Jaszbereny	Hungary	250		_	_	
		Bogota	Colombia	1	L	_	_	
		Julich	Germany (W)	100	/	_	_	_
		Wertachtal	Germany (W)	500	/	_	_	_
		Luanda	Angola	100	/	_	_	
		Cyclops	Malta	250	/		_	_
		Peking	China Rep					
5.965	50.29	London	UK	100/500	L	_	_	_
		Huanuni	Bolivia	10	L	_	_	_
		P Alegre	Brasil	7.5	L	_	_	_
		Kajang	Malaysia	100	L	_	_	_
		Ismaning	Germany (W)	100	L	_	_	_
		Mt Carlo	Monaco	100	L	_	_	_
		Tanger	Morocco	35	L	_	_	_
		Malolos	Philippines	7.5	_	_	_	_
		Moskva	USSR	50	L	_	_	_
		Armavir	USSR	100		_	_	
		Rhodos	Greece	50	L	_	_	_
		Diosd	Hungary	100	L	_	_	_
		S Pedro Sula	Honduras Rep	1	/		_	_
		Benin	Niger	10	_	_		
		Jos	Niger	10	/	_	_	
		Rangoon	Burma	50	L	_	_	_
		Wavre	Belgium	100	/	_	_	_

MHz	Metres	Station	Country	kW	M	M	S	N
		Domingo	Dominican Rep					
5.970	50.25	Bandjarmasin	Indonesia	10	∠	—	—	—
		Bogota	Colombia	1	∠	—	—	—
		Tula	USSR	240	—			
		Alma Ata	USSR	100	∠	—	—	—
		Moskva	USSR	100	∠	—	—	—
		Komsomolskamur	USSR	50		—	—	
		Nikolaevskamur	USSR	50	∠		—	
		Tchita	USSR	240			—	
		Gauhati	India	10	∠	—	—	—
		Aden	Aden	100				
		Arganda	Spain	100	—	—	—	—
		Holzkirchen	Germany (W)	10	—			
		Biblis	Germany (W)	100	∠	—	—	—
		Lisbonne	Portugal	100				—
		Diosd	Hungary	100	—			
		Shepparton	Australia	10			—	—
		Schwarzenburg	Switzerland	150			—	—
5.975	50.21	London	UK	100/500	∠	—	—	—
		Florianapolis	Brasil	1	∠	—	—	—
		Kyung San	Korea (S)	10	∠	—	—	—
		Cochabamba	Bolivia	1	∠	—	—	—
		Villarrica	Paraguay	3	∠	—	—	—
		Gwelo	Rhodesia	10/100		—	—	—
		Komsomolskamur	USSR	50		—		
		Tachkent	USSR	100	∠			—
		Minsk	Bielorussia	100	∠			—
		Peking	China Rep					
5.980	50.17	Beyrouth	Lebanon	100		—	—	—
		Meyerton	South Africa	250/500	∠	—	—	—
		Godthaab	Greenland	1	∠	—	—	—
		Riazan	USSR	240	∠	—	—	—
		Tbilisi	USSR	240	∠		—	
		Irkutsk	USSR	100	∠			—
		Quetta	Pakistan	10	∠	—	—	—
		Goderich	Sierra Leone	10	∠	—	—	—
		Waterloo	Sierra Leone	250	∠	—	—	—
		Redwood City	USA	30/250	—		—	—
		Linares	Mexico	0.5	∠	—	—	—
		Taipei	China Nat					
		Diosd	Hungary	100	∠	—	—	—
		Jaszbereny	Hungary	250	∠	—	—	—
		Szekesfehervar	Hungary	20				
		Medellin	Colombia	10	∠	—	—	—
		Cyclops	Malta	250	/			
		RFE						
		Sparendum	Guyana	2	/			
		Jakarta	Indonesia					

MHz	Metres	Station	Country	kW	M	M	S	N
5.985	50.13	Dar es Salaam	Tanzania	100				
		Lomas Mirador	Argentina	1	∠	_	_	_
		Lisbonne	Portugal	100/250	_			_
		Holzkirchen	Germany (W)	10		_		
		Lampertheim	Germany (W)	100	∠		_	
		Biblis	Germany (W)	100	∠	_	_	_
		Mexico	Mexico	10	∠	_	_	_
		Khabarovsk	USSR	120	_			
		Scituate	USA	50/100	∠	_	_	_
		Rabaul	Papua New Guinea	10	∠	_	_	_
		Kavalla	Greece	250			_	_
		Rangoon	Burma	50	∠	_	_	_
		Allouis	France	100	∠	_	_	_
		Tunja	Colombia	10	∠	_	_	_
5.990	50.08	London	UK	100/250	∠	_	_	_
		Limassol	Cyprus	10	∠			_
		Bhopal	India	10	∠	_	_	_
		Ejura	Ghana	10	∠	_	_	_
		Serpukhov	USSR	100	∠			_
		Nikolaevskamur	USSR	50	∠	_	_	_
		Roma	Italy	60/100	∠	_	_	_
		Rio de Janeiro	Brasil	7.5/10	∠	_	_	_
		Menado	Indonesia	10	∠	_	_	_
		Bucuresti	Roumania	120	∠	_	_	_
		Tokyo Yamata	Japan	100	_			
		Sackville	Canada	50/250	_	_	_	_
		Allouis	France	100	_	_	_	_
		Cyclops	Malta	250	_			
		Arganda	Spain	100	_	_	_	_
		Gedja	Ethiopia	100	/	_	_	_
		Peking	China Rep					
		V. of Kurdistan						
5.995	50.04	London	UK	100	∠	_	_	_
		Bamako	Malawi	50		_	_	
		Greenville	USA	250/500	∠	_	_	_
		Mbandaka	Zaire	10	∠	_	_	_
		Panama	Panama	4	∠	_	_	_
		Ft de France	Martinique	4	∠	_	_	_
		Warszawa	Poland	8/10	_	_	_	_
		Limbe	Malawi	20	∠	_	_	_
		Sucre	Bolivia	1	_	_	_	_
		Lyndhurst	Australia	10	∠	_	_	_
		Julich	Germany (W)	100	∠	_	_	_
		Poro	Philippines	35	_			
		S M Galeria	Vatican	100	∠	_	_	_
		P J Caballero	Paraguay	2	∠	_	_	_
		S Pedro Sula	Honduras Rep	1	/		_	_
		Tula	USSR	50		_	_	

MHz	Metres	Station	Country	kW	M	M	S	N
		Allouis	France	100	∠	_	_	_
		Pereira	Colombia	1	∠	_	_	_
6.000	50.00	Montserrat	Br W Indies	15	_	_	_	_
		Singapore	Singapore	10/50	∠	_	_	_
		Belo Horizonte	Brasil	1/25	∠	_	_	_
		Montevideo	Uruguay	5	∠	_	_	_
		Innsbruck	Austria	10	_	_	_	_
		Tchita	USSR	100	∠	_	_	_
		Moskva	USSR	240	∠	_	_	_
		Kabul	Afghanistan	50/100	∠	_	_	_
		Diriyya	Saudi Arabia	50	∠	_	_	_
		Jaszbereny	Hungary	250	∠	_	_	_
		Szekesfehervar	Hungary	20	_	_	_	
		Cyclops	Malta	250	∠	_	_	_
		Arganda	Spain	100	_	_	_	_
6.005	49.96	London	UK	250	/			
		Ascension	Ascension	126/250	∠	_	_	_
		Buea	Cameroon	4	∠	_	_	_
		Ekala	Ceylon	10	∠	_	_	_
		La Paz	Bolivia	10	∠	_	_	_
		Marhubi	Zanzibar	3.5	_	_	_	_
		Montreal	Canada	0.5	∠	_	_	_
		Ismaning	Germany (W)	20	∠	_	_	_
		Berlin (RIAS)	Germany (W)	20	_	_	_	_
		S Jose	Costa Rica	1	∠	_	_	_
		Voronej	USSR	100	_	_	_	_
		Matsuyama	Japan	0.6	∠	_	_	_
		Carnarvon	Australia	250	∠	_	_	_
		Fredrikstad (Relay of Yemen Arab Rep)	Norway	100				_
6.010	49.92	London	UK	100/250	∠	_	_	_
		Kranji	Singapore	250	_	_	_	
		Limassol	Cyprus	100	∠	_	_	
		Wavre	Belgium	50/100	∠	_	_	_
		Calcutta	India	10	_	_	_	_
		Mexico	Mexico	5	∠	_	_	_
		Montevideo	Uruguay	10	∠	_	_	_
		Moskva	USSR	240/500	∠	_	_	_
		Krasnoiarsk	USSR	240	∠	_	_	_
		Delano	USA	100				_
		Allouis	France	100/500	∠	_	_	_
		Managua	Nicaragua	0.1	∠	_	_	_
		Sines	Portugal	250				_
		Roma	Italy	60/100			_	_
		K Wusterhausen	Germany (E)	100	∠			
		Nauen	Germany (E)	100			_	_

MHz	Metres	Station	Country	kW	M	M	S	N
		Islamabad	Pakistan	100	/		—	—
		Holzkirchen	Germany (W)	10	/			
		Biblis	Germany (W)					
		Tinang	Philippines	250		—	—	—
		Meyerton	South Africa	100	—			
		Pereira	Colombia	10	∟	—	—	—
		Quito	Ecuador	100	—			
6.015	49.88	London	UK	250	/			—
		Abidjan	Ivory Coast	100	∟	—	—	—
		Asuncion	Paraguay	1	∟	—	—	—
		Animas	Bolivia	5	∟	—	—	—
		Recife	Brasil	10	∟	—	—	—
		Rhodos	Greece	50	∟	—	—	—
		Dixon	USA	250	—		—	—
		Orcha	Bielorussia	100				—
		Tanger	Morocco	50/100	∟	—		—
		Velkekostolany	Czechoslovakia	120	∟	—	—	—
		Sines	Portugal	250	∟		—	—
		Delhi	India	100	∟	—	—	—
		Gedja	Ethiopia	100	—	—	—	—
		Wien	Austria	100	—	—	—	—
		Fredrikstad	Norway	100/250	∟	—	—	—
		Tumaco	Colombia	1	∟	—	—	—
		S M Galeria	Vatican	100	∟	—	—	—
		Wavre	Belgium	250	∟		—	—
6.020	49.85	Limassol	Cyprus	100			—	—
		Lopik	Netherlands	10/100	∟	—	—	—
		Khabarovsk	USSR	50	∟	—	—	—
		Kiev	Ukraine	50	∟	—	—	—
		Simla	India	2.5	∟	—	—	—
		Delhi	India	100	∟	—	—	—
		Vera Cruz	Mexico	5	∟	—	—	—
		Greenville	USA	250/500	∟	—	—	—
		Bonaire Noord	Neth Antilles	300	∟	—	—	—
		Talata Volon	Malagasy Rep	300	∟	—	—	—
		Gwelo	Rhodesia	20/100	—	—		
		Tegucigalpa	Honduras Rep	0.5	/		—	—
		Bogota	Colombia	10	∟		—	—
6.025	49.79	Asuncion	Paraguay	10/100	∟	—	—	—
		Kajang	Malaysia	50	∟	—	—	—
		S Gabriel	Portugal	100	∟	—	—	—
		S Pedromacoris	Dominican Rep	0.1	∟	—	—	—
		Enugu	Niger	10	∟	—	—	—
		Jaszbereny	Hungary	250	∟	—	—	—
		Diosd	Hungary	100	/			—
		La Paz	Bolivia	10	∟	—	—	—
		Tachkent	USSR	50	—		—	—
		Cyclops	Malta	250	∟	—	—	—
		Beira	Mozambique	10	∟			

MHz	Metres	Station	Country	kW	M	M	S	N
		National Voice of Iran						
6.030	49.75	London	UK	100/250	∟	_	_	_
		Masirah	Oman	200	/			
		Greenville	USA	50	∟	_	_	_
		Cincinnati	USA	175	_	_	_	_
		Tokyo Yamata	Japan	20/50	∟	_	_	_
		Bogota	Colombia	25	∟	_	_	_
		Calgary	Canada	0.1	∟	_	_	_
		Moskva	USSR	100	∟	_	_	_
		Komsomolskamur	USSR	240				_
		Simferopol	Ukraine	240	∟			
		Muehlacker	Germany (W)	20	∟	_	_	_
		Bocaue	Philippines	25	∟	_	_	_
		Kampala	Uganda	250				_
		Kimjae	Korea (S)	100				_
		Islamabad	Pakistan	10/100	_	_	_	_
		Franceville	Gabon Rep	500				_
6.035	49.71	Warszawa	Poland	60	∟	_	_	_
		Bombay	India	100	_	_	_	_
		La Paz	Bolivia	10	∟	_	_	_
		Careysburg	Liberia	250	∟	_	_	_
		Mt Carlo	Monaco	100	∟	_	_	_
		Montevideo	Uruguay	1	∟	_	_	_
		Vladivostock	USSR	100	_	_	_	_
		Rio de Janeiro	Brasil	10	∟	_	_	_
		Carnarvon	Australia	250	∟	_	_	_
		Tegucigalpa	Honduras Rep	0.5	/			
6.040	49.67	London	UK	250	∟	_	_	_
		Antigua	Br W Indies	250	∟	_	_	_
		Dubai	United Arab Emirates	10	∟	_	_	_
		Ibague	Colombia	10	∟	_	_	_
		S Jose	Costa Rica	1	∟	_	_	_
		Kazan	USSR	100	_			
		Simferopol	Ukraine	100	_			
		Alotau	Papua New Guinea	10	∟	_	_	_
		Nauen	Germany (E)	500	∟	_	_	_
		Cincinnati	USA	175	_		_	_
		Allouis	France	100	∟	_	_	_
		Jaszbereny	Hungary	250	∟	_	_	_
		Szekesfehervar	Hungary	20	∟	_	_	_
		Ismaning	Germany (W)	100	/		_	_
		Julich	Germany (W)	100			_	_
6.045	49.63	Athinai	Greece	5	∟	_	_	_
		Curityba	Brasil	7.5	∟	_	_	_
		David	Panama	1	∟	_	_	_
		Careysburg	Liberia	50/250	∟	_		_

64

MHz	Metres	Station	Country	kW	M	M	S	N
		Jakarta	Indonesia	100	∠	_	_	_
		Montevideo	Uruguay	2.5	∠	_	_	_
		Moskva	USSR	240	∠	_	_	_
		Novosibirsk	USSR	100	_			_
		Tchita	USSR	240/500	_			_
		S Luis Potosi	Mexico	0.3	∠	_	_	_
		Lopik	Netherlands	10/100	∠	_	_	_
		Lyndhurst	Australia	10	∠	_	_	_
		Arganda	Spain	100	∠	_	_	_
		Las Mesas	Canary Is	50	_		_	_
		Schwarzenburg	Switzerland	150	∠		_	_
		Tungsheng	Mongolian Rep					
		Tambacounda	Senegal	4	/			
		Sackville	Canada	50	/			
6.050	49.59	London	UK	100/250	∠	_	_	_
		Limassol	Cyprus	100	∠	_		_
		Kranji	Singapore	250	∠	_		_
		Sibu	Malaysia	10		_		
		Ibadan	Niger	10	∠		_	_
		Delhi	India	20	∠	_	_	_
		Irkutsk	USSR	50	∠	_	_	_
		Maputo	Mozambique	10/20	∠	_	_	_
		Quito	Ecuador	100	∠	_	_	_
		Roma	Italy	100	∠			_
		S Jose	Costa Rica	1	∠	_		_
		Noblejas	Spain	350	_	_	_	_
		Tegucigalpa	Honduras Rep	0.5	/		_	_
		Sanaa	Yemen	50				
6.055	49.55	Ascension	Ascension	125/250	_			
		Antigua	Br W Indies	250	_		_	
		Greenville	USA	50	∠	_	_	_
		Dixon	USA	250	/	_		_
		Kigali	Rwanda	50	∠	_	_	_
		Sulaibiyah	Kuwait	250	∠	_	_	_
		Melo	Uruguay	5	∠	_	_	_
		Erevan	USSR	100	∠			_
		Tchita	USSR	50				_
		Tallin	USSR	240			_	
		Starobelsk	Ukraine	100	∠	_	_	_
		La Paz	Bolivia	100	∠	_	_	_
		Velkekostolany	Czechoslovakia	120	∠	_	_	_
		Praha	Czechoslovakia	120/400	∠	_	_	_
		S Paulo	Brasil	7.5	∠	_	_	_
		Tokyo Nagara	Japan	50	∠	_	_	_
		Cali	Colombia	5	∠	_	_	_
6.060	49.50	London	UK	250	_	_	_	_
		Gral Pacheo	Argentina	50	∠	_	_	_

MHz	Metres	Station	Country	kW	M	M	S	N
		Caltanissetta	Italy	25	L	_	_	_
		Roma	Italy	50	L	_	_	_
		Alma Ata	USSR	100		_	_	
		Tbilisi	USSR	100	L		_	
		Komsomolskamur	USSR	240	L	_	_	_
		Krasnoiarsk	USSR	100	_			
		Habana	Cuba	10/50	L	_	_	_
		Lusaka	Zambia	20	L	_	_	_
		Santiago	Dominican Rep	0.5	L	_	_	_
		Kavalla	Greece	250	L	_	_	_
		Mira	Malaysia	10		_		
		Ismaning	Germany (W)	100	L	_	_	_
		Tanger	Morocco	35/100	_	_	_	_
		Tegucigalpa	Honduras Rep	1	/		_	_
		Jaszbereny	Hungary	250		_	_	
		Shepparton	Australia	100	/			
6.065	49.46	London	UK	250	_		_	
		Kranji	Singapore	250	/		_	
		Rio de Janeiro	Brasil	7.5	L	_	_	_
		Brasilia	Brasil	10		_	_	_
		Hoerby	Sweden	350	L	_	_	_
		Karlsborg	Sweden	350	L	_	_	_
		Kazan	USSR	100	L			
		Armavir	USSR	100	L	_	_	_
		Serpukhov	USSR	100		_		
		Texmelucan	Mexico	1	L	_	_	_
		Kohima	India	2		_	_	_
		Bogota	Colombia	2	L	_	_	_
		Wertachtal	Germany (W)	500	L	_	_	_
		Julich	Germany (W)	100	L	_	_	_
		Arganda	Spain	100	_	_	_	_
		Noblejas	Spain	100	/			
		Sackville	Canada	50/250	L	_	_	_
		Wavre	Belgium	100	_			
6.070	49.42	London	UK	250	/			_
		Limassol	Cyprus	100	L		_	_
		Tema	Ghana	100	L	_	_	_
		Bangkok	Thailand	10	L	_	_	_
		Sofia	Bulgaria	50/100	L	_	_	_
		Jajapura	Indonesia	0.5/20	L	_	_	_
		Toronto	Canada	1	L	_	_	_
		Oruro	Bolivia	5	L	_	_	_
		Khabarovsk	USSR	100/150		_	_	_
		Tula	USSR	50		_	_	
		Leipzig	Germany (E)	10	L			
		K Wusterhausen	Germany (E)	100		_	_	
		Nauen	Germany (E)	500	_			_
		Rhodos	Greece	50	L	_	_	_
		Manzini	Swaziland	25		_	_	_

MHz	Metres	Station	Country	kW	M	M	S	N
		Quito	Ecuador	100	/		_	_
		Biblis	Germany (W)	100	_		_	
		Lisbonne	Portugal	100				
		Arganda	Spain	100	/			_
6.075	49.38	Ekala	Ceylon	10	L	_	_	_
		Julich	Germany (W)	100	L	_	_	_
		Wertachtal	Germany (W)	500	L	_	_	_
		Roma	Italy	60	L			_
		Montevideo	Uruguay	2.5	L	_	_	_
		Santiago	Dominican Rep	0.5	L	_	_	_
		Bogota	Colombia	10/50	L	_	_	_
		Sverdlovsk	USSR	50				_
		Tula	USSR	50			_	
		Volgograd	USSR	100			_	_
		Quito	Ecuador	100		_	_	_
		Noblejas	Spain	350	_			
		Delhi	India	20	_	_	_	
		Monrovia	Liberia	50	L	_	_	_
		S Barbara	Honduras Rep	1	/		_	_
6.080	49.34	London	UK	250	L	_	_	_
		Kranji	Singapore	250	L	_	_	_
		K Wusterhausen	Germany (E)	100	L			
		Nauen	Germany (E)	50	L	_		_
		Frunze	USSR	240	/			_
		Moskva	USSR	100	L		_	
		Komsomolskamur	USSR	50	L	_	_	_
		Vancouver	Canada	0.1	L	_	_	_
		Tanger	Morocco	35/100				_
		Riyadh	Saudi Arabia	350	L	_	_	_
		Bangkok	Thailand	2	L	_	_	_
		Daru	Papua New Guinea	10	L	_	_	_
		Wavre	Belgium	100	/	_	_	_
		Greenville	USA	50/500	L	_	_	_
		Alger	Algeria	100	L	_	_	_
		Tambacounda	Senegal	4	_	_	_	_
		Peshawar	Pakistan	10	L	_	_	_
		Hailar	Mongolian Rep					
		Diosd	Hungary	100	_			
		Jaszbereny	Hungary	250	_	_	_	
		Catavi	Bolivia					
		Shepparton	Australia	10				_
6.085	49.30	London	UK	100	_		_	
		Limassol	Cyprus	100	_		_	_
		Antigua	Br W Indies	250	L	_	_	_
		Lopik	Netherlands	100	L	_	_	_
		Kisangani	Zaire	10	L	_	_	_
		Delhi	India	100	_	_	_	_
		Madras	India	100	_	_	_	_

MHz	Metres	Station	Country	kW	M	M	S	N
		Ismaning	Germany (E)	10	∟	_	_	_
		Wertachtal	Germany (E)	500	∟	_	_	_
		Recife	Brasil	15	∟	_	_	_
		Tbilisi	USSR	100	∟		_	
		Tallin	USSR	50	∟	_	_	_
		Kiev	Ukraine	50		_	_	
		Sackville	Canada	125/250	∟	_	_	_
		Noblejas	Spain	350	∟	_	_	_
		Riyadh	Saudi Arabia	350	∟	_	_	_
		Tegucigalpa	Honduras Rep	1	/		_	_
		Sofia	Bulgaria	50/100	∟	_	_	_
		Fredrikstad	Norway	100			_	
6.090	49.26	Hurlingham	Argentina	30	∟	_	_	_
		Cd Mante	Mexico	1	∟	_	_	_
		Simferopol	Ukraine	240		_	_	
		Orenburg	USSR	50	∟	_	_	_
		Nikolaevskamur	USSR	100	∟			
		Irkutsk	USSR	50	∟	_	_	_
		Junglinster	Luxembourg	500	∟	_	_	_
		S Domingo	Dominican Rep	7.5	∟	_	_	_
		Jaji	Niger	10	∟	_	_	_
		Sydney	Australia	2	∟	_	_	_
		Beira	Mozambique	25	∟	_	_	_
		Kavalla	Greece	250	∟	_	_	_
		Las Mesas	Canary Is	50		_	_	_
		V of Kampuchea People						
6.095	49.22	Ismaning	Germany (E)	100	_			
		Warszawa	Poland	30	∟	_	_	_
		Salman Pack	Iraq	50	_	_	_	_
		S Paulo	Brasil	25	∟	_	_	_
		Mogadiscio	Somalia	50	∟	_	_	_
		Espinal	Colombia	10	∟	_	_	_
		Kavalla	Greece	250		_	_	
		Dixon	USA	100	_	_	_	_
		Quito	Ecuador	100	∟	_	_	_
		Tanger	Morocco	35/50	∟	_	_	_
		Serpukhov	USSR	100	∟		_	
		Tegucigalpa	Honduras Rep	1	/		_	_
		Phnom Penh	Cambodia					
6.100	49.18	Belgrade	Yugoslavia	100	∟	_	_	_
		Wertachtal	Germany (W)	500	∟	_	_	_
		Kajang	Malaysia	100	∟	_	_	_
		Nueva Segovia	Nicaragua	0.4	_			
		Ocotal	Nicaragua	0.4	/	_	_	_
		Kursk	USSR	50	∟	_	_	
		Kaunas	USSR	50	∟	_	_	
		Petropavlo Kam	USSR	240	∟	_	_	_
		Vinnitsa	Ukraine	100			_	

MHz	Metres	Station	Country	kW	M	M	S	N
		Maiduguri	Niger	10	L	_	_	_
		Kurseong	India	20	_	_	_	_
		Hoerby	Sweden	350	_			
		Cyclops	Malta	250	L	_	_	_
		Arganda	Spain	100	L	_	_	_
		Tanger	Morocco	50	_	_		
		Rabat	Morocco	50	/	_		
		Tripoli	Libya					
		Kathmandu	Nepal	100	L		_	_
		R of the Patriots						
		Caracas	Venezuela					
6.105	49.14	Fortaleza	Brasil	5	L	_	_	_
		Medellin	Colombia	10	L	_	_	_
		Merida	Mexico	1	L	_	_	_
		Armavir	USSR	100	_			_
		Kalinin	USSR	120	L			_
		Duchanbe	USSR	100	L			_
		Delhi	India	100	_	_	_	_
		Dar es Salaam	Tanzania	50	_	_	_	_
		Cite Vatican	Vatican	80	L	_	_	_
		Wellington	New Zealand	7.5	L	_	_	_
		Diosd	Hungary	100	/			_
		Jaszbereny	Hungary	100	_	_	_	_
		Szekesfehervar	Hungary	20	/	_	_	_
		Biblis	Germany (W)	100	L	_	_	_
		Holzkirchen	Germany (W)	10	_	_	_	
		Lampertheim	Germany (W)	100	L			_
		Lisbonne	Portugal	250	_			_
		Sackville	Canada	50/250				_
6.110	49.10	London	UK	250	L	_	_	_
		Ascuncion	Paraguay	3	L	_	_	_
		Baku	USSR	50	L	_	_	_
		Kazan	USSR	50				_
		S Domingo	Dominican Rep	1	L	_	_	_
		Szekesfehervar	Hungary	20	L	_	_	_
		Diosd	Hungary	100	/			_
		Jaszbereny	Hungary	250	_	_		
		Srinagar	India	7.5	_	_	_	_
		Tinang	Philippines	50/250	L	_	_	_
		Ismaning	Germany (W)	100	L	_	_	_
		Comayaguela	Honduras Rep	0.5	/	_	_	
		Tanger	Morocco					
6.115	49.06	London	UK	100	_			_
		K Wusterhausen	Germany (E)	5	L	_	_	_
		Hermosillo	Mexico	1	L	_	_	_
		Khabarovsk	USSR	50/240	L	_	_	_
		Lvov	Ukraine	240	L	_	_	_
		Montevideo	Uruguay	5	L	_	_	_
		Maputo	Mozambique	25	L	_	_	_

MHz	Metres	Station	Country	kW	M	M	S	N
		Biblis	Germany (W)	100	∠	—	—	—
		Holzkirchen	Germany (W)	10	∠	—	—	—
		Lisbonne	Portugal	250			—	—
		Sines	Portugal	250	∠		—	—
		Brazzaville	Congo	50	∠	—	—	—
		Rio de Janeiro	Brasil	10	∠	—	—	—
		Karachi	Pakistan	50	∠	—	—	—
		Tokyo Nagara	Japan	10	∠	—	—	—
		Villavicencio	Colombia	1	∠	—	—	—
		Szekesfehervar	Hungary	20				
6.120	49.02	London	UK	100/250	—	—	—	—
		Limassol	Cyprus	7.5	∠	—	—	—
		Hyderabad	India	10	—	—	—	—
		Delhi	India	10/100	∠	—	—	—
		S Fernando	Argentina	10	∠	—	—	—
		Julich	Germany (W)	100	∠	—	—	—
		Wertachtal	Germany (W)	500	∠	—	—	—
		Bluefields	Nicaragua	1	∠	—	—	—
		Bocaue	Philippines	50	∠	—	—	—
		Pori	Finland	15/250	∠	—	—	—
		Tapachula	Mexico	0.5	∠	—	—	—
		Armavir	USSR	240	∠	—	—	—
		Moskva	USSR	100	—			—
		Novosibirsk	USSR	100	—			—
		Duchanbe	USSR	120	∠			—
		Santiago	Dominican Rep	0.5	∠	—	—	—
		Noblejas	Spain	350	∠	—	—	—
		Surabaja	Indonesia	10	∠	—	—	—
		Sackville	Canada	50/250	∠	—	—	—
		Bata	Guinea	50	∠	—	—	—
		Shepparton	Australia	50	/			
6.125	48.98	London	UK	100/500	∠	—	—	—
		Limassol	Cyprus					
		Greenville	USA	250/500	∠	—	—	—
		Dixon	USA	250	—		—	—
		Cincinnati	USA	250	∠	—	—	—
		Kananga	Zaire	10	∠	—	—	—
		Montevideo	Uruguay	10	∠	—	—	—
		La Paz	Bolivia	1	∠	—	—	—
		S Paulo	Brasil	10	∠	—	—	—
		Achkhabad	USSR	100	∠	—	—	—
		Lvov	Ukraine	240				—
		Krasnoiarsk	USSR	100	∠	—	—	—
		Bogota	Colombia	1	∠	—	—	—
		Jaszbereny	Hungary	250			—	—
		S Pedro Sula	Honduras Rep	1	/			—
6.130	48.94	London	UK	100	∠	—	—	—
		Limassol	Cyprus	100			—	—
		Ekala	Ceylon	10	∠	—	—	—

70

MHz	Metres	Station	Country	kW	M	M	S	N
		Julich	Germany (W)	100	∠	_	_	_
		Wertachtal	Germany (W)	500	∠	_	_	_
		Halifax	Canada	0.5	∠	_	_	_
		Armavir	USSR	100	/			
		Moskva	USSR	240	_	_	_	_
		Novosibirsk	USSR	100	∠	_	_	_
		Simferopol	Ukraine	100	∠	_		_
		Greenville	USA	250	∠	_	_	_
		S Domingo	Dominican Rep	0.3	∠	_	_	_
		Tema	Ghana	100	∠	_	_	_
		Kumamoto	Japan	1	∠	_	_	_
		Vientiane	Laos	10	∠	_	_	_
		Karlsborg	Sweden	350	_			
		Quito	Ecuador	100	∠	_	_	_
		Rawalpindi	Pakistan	10	/	_	_	_
		Gauhati	India	10	_	_	_	_
		Tanger	Morocco	100				
6.135	48.90	Montserrat	Br W Indies	15	_	_	_	_
		Baku	USSR	100	∠	_	_	_
		Alma Ata	USSR	50	∠	_	_	
		Khabarovsk	USSR	100	_	_		
		Biblis	Germany (W)	100	/			
		Holzkirchen	Germany (W)	10/50	_			
		Lisbonne	Portugal	250	_		_	
		Warszawa	Poland	40/100	∠	_	_	_
		Papeete	Tahiti	4/20	∠	_	_	_
		P Alegre	Brasil	7.5	∠	_	_	_
		Suwon	Korea (S)	10	∠	_	_	_
		Samarinda	Indonesia	7.5	∠	_	_	_
		S Cruz	Bolivia	1	∠	_	_	_
		Tananarive	Malagasy Rep	30	/	_	_	_
		Schwarzenburg	Switzerland	150	∠	_	_	_
		Santiago	Chile	100				
		Concepcion	Chile	10	∠			_
		La Ceiba	Honduras Rep	0.5	/	_	_	
		Greenville	USA	50	∠	_	_	
6.140	48.86	London	UK	100/500	∠	_	_	_
		Masirah	Oman	100/200				_
		Ascension	Ascension	250	_	_		
		Limassol	Cyprus	100	_	_		
		Chihuahua	Mexico	0.3	∠	_	_	_
		Voronej	USSR	100	∠	_	_	_
		Kiev	Ukraine	100	∠	_	_	_
		Arganda	Spain	100	∠	_	_	_
		Montevideo	Uruguay	10	∠	_	_	_
		Perth	Australia	10	∠	_	_	_
		Bujumbura	Burundi	10/25	∠	_	_	_
		Wewak	Papua New Guinea	10	∠	_	_	_

MHz	Metres	Station	Country	kW	M	M	S	N
		Kavalla	Greece	250	∟	—	—	—
		Athinai	Greece	100	∟	—	—	—
		Ranchi	India	2	∟	—	—	—
		Delhi	India	100	—	—	—	—
		Sackville	Canada	50/250	∟	—	—	—
		Bogota	Colombia	5	∟	—	—	—
6.145	48.82	Montserrat	Br W Indies	15/50	—	—	—	—
		Delhi	India	20/100	∟	—	—	—
		Julich	Germany (W)	100	∟	—	—	—
		Wertachtal	Germany (W)	500	∟	—	—	—
		Moskva	USSR	100	∟	—	—	—
		Khabrovsk	USSR	100	∟			
		Starobelsk	Ukraine	240	∟			—
		Tlaxiaco	Mexico	0.3	∟	—	—	—
		Delano	USA	250	—		—	—
		Allouis	France	500	∟	—	—	—
		Calabar	Niger	10	∟	—	—	—
		Alger	Algeria	50	∟	—	—	—
		Tarija	Bolivia	1	∟	—	—	—
		Juticalpa	Hungary	0.5	/		—	—
		Dacca	Bangladesh	7.5	—	—	—	
		Popayan	Colombia	1	∟	—	—	—
6.150	48.78	London	UK	100/500	∟	—	—	—
		Limassol	Cyprus	100	—		—	
		Belgrade	Yugoslavia	10/50	∟	—	—	—
		Bucuresti	Roumania	120	∟	—	—	—
		Ismaning	Germany (W)	100	∟	—	—	—
		Lyndhurst	Australia	10	∟	—	—	—
		S Jose	Costa Rica	1	∟	—	—	—
		Omdurman	Sudan	20				
		Kavalla	Greece	250	∟	—	—	—
		Ekala	Ceylon	10	∟	—	—	—
		Male	Maldives	2.7	—	—	—	—
		Serpukhov	Ukraine	50	/			—
		Komsomolskamur	USSR	240	—			—
		Kazan	USSR	100		—		
		Delhi	India	20	—	—	—	—
		Santiago	Chile	5/100	∟			—
		Sackville	Canada	50	—			
		Neiva	Colombia	1	∟	—	—	—
6.155	48.74	Limassol	Cyprus	100	/	—		—
		Salman Pack	Iraq	100				
		La Paz	Bolivia	1	∟	—	—	—
		S Gabriel	Portugal	50	∟	—	—	—
		Krasnoiarsk	USSR	50	∟			—
		Nikolaevskamur	USSR		∟	—	—	—
		Montevideo	Uruguay	10	∟	—	—	—
		Singapore	Singapore	50	∟	—	—	—
		Tokyo	Japan	10	∟	—	—	—

MHz	Metres	Station	Country	kW	M	M	S	N
		Wien	Austria	100	∠	-	-	-
		Warszawa	Poland	100	/	-	-	-
		Bucuresti	Roumania	18/120	∠	-	-	-
		Conakry	Guinea	18/100	∠	-	-	-
		Scituate	USA	50/100		-	-		-
		Cincinnati	USA	175		-	-		-
		Salvador	Brasil	10	∠	-	-	-
		Dacca	Bangladesh	7.5	/			-
6.160	48.70	London	UK	250/500	∠	-	-	-
		Delhi	India	100	∠	-	-	-
		Sofia	Bulgaria	50/120	∠	-	-	-
		Bogota	Colombia	10	∠	-	-	-
		St Johns	Canada	0.3	∠	-	-	-
		Vancouver	Canada	0.5	∠	-	-	-
		Malargue	Argentina	0.3/3	∠	-	-	-
		Wavre	Belgium	50/100				
		Moskva	USSR	100	∠	-	-	-
		Alger	Algeria	50	∠	-	-	-
		Kigali	Rwanda	250	∠	-	-	-
		Jaszbereny	Hungary	250		-	-	-
		Arganda	Spain	100	/			
6.165	48.66	Lenk	Switzerland	250	∠	-	-	-
		Vladivostock	USSR	100	/	-	-	-
		Kiev	Ukraine	100	∠	-	-	-
		Lusaka	Zambia	20	∠	-	-	-
		Mexico	Mexico	10	∠	-	-	-
		S Paulo	Brasil	7.5	∠	-	-	-
		Bonaire Noord	Neth Antilles	300	∠	-	-	-
		Tegucigalpa	Honduras Rep	0.5/1	/		-	-
		Ho Chi Minh City	Vietnam					
		Jaszbereny	Hungary	250		-	-	-
		N Djamena	Chad	100				
6.170	48.62	London	UK	100				-
		Holzkirchen	Germany (W)	10	-			-
		Biblis	Germany (W)	100	/			-
		Lampertheim	Germany (W)	20/100		-	-	
		Cayenne	Guyana Fr	4	∠	-	-	-
		Lucknow	India	10	∠	-	-	-
		Marulas	Philippines	10	∠	-	-	-
		Montevideo	Uruguay	1	∠	-	-	-
		Jigulevsk	USSR	240	∠			-
		Armavir	USSR	100	∠			-
		Starobelsk	Ukraine	240	∠			-
		Tanger	Morocco	50/100		-		-
		Tananarive	Malagasy Rep	30	/	-	-	-
		Kavalla	Greece	250		-	-	-
		Abu Zaabal	Egypt	60/100				-
		Caracas	Venezuela	10	∠	-	-	-

73

MHz	Metres	Station	Country	kW	M	M	S	N
		Florencia	Colombia	2.5	∟	_	_	_
		Kimjae	Korea (S)	100	∟	_	_	_
6.175	48.58	Antigua	Br W Indies	125/250	∟	_	_	_
		Belo Horizonte	Brasil	10	/	_	_	_
		Kajang	Malaysia	100	∟	_	_	_
		Khabarovsk	USSR	100	_	_		
		Kazan	USSR	120	∟	_		
		Krasnoiarsk	USSR	100				_
		Moskva	USSR	100	_	_	_	
		Vinnitsa	Ukraine	240/500	∟			_
		Allouis	France	100	∟	_	_	_
		Hiroshima	Japan	1/5	∟	_	_	_
		S Cruz	Bolivia	5	∟	_	_	_
		Kaduna	Niger	10/20	∟	_	_	_
		Sackville	Canada	50/250	_	_	_	
		Seeb	Oman	50	∟	_	_	_
			China Rep					
		Luanda	Angola	100	/		_	_
		Koebenhaven	Denmark	50	_			
		Faro Caribe	Costa Rica	2.5	/			_
6.180	48.54	London	UK	250	∟	_	_	_
		Limassol	Cyprus	100	∟	_	_	_
		Mendoza	Argentina	10	∟	_	_	_
		Alma Ata	USSR	100	/	_	_	_
		Tula	USSR	100	∟	_	_	_
		Careysburg	Liberia	50	∟	_	_	_
		Bogota	Colombia	25	/	_	_	_
		Fredrikstad	Norway	100	∟	_		
		Bucuresti	Roumania	18/250	∟	_	_	_
		Ziguinchor	Senegal	4	_	_	_	_
		Tambacounda	Senegal	4	/	_	_	_
		Tanger	Morocco	100	_			
		Guatemala City	Guatemala					
		Islamabad	Pakistan	100	/	_	_	_
		Dacca	Bangladesh	250	/	_	_	_
		Tirana	Albania					
6.185	48.50	London	UK	250	_			
		Gedja	Ethiopia	100	_			
		Ekala	Ceylon	10	∟	_	_	_
		Julich	Germany (W)	100	∟	_	_	_
		Wertachtal	Germany (W)	500	∟	_	_	_
		La Paz	Bolivia	1	∟	_	_	_
		Tripoli	Libya	100	∟	_	_	
		Manokwari	Indonesia	1/10	∟	_	_	_
		Mexico	Mexico	1	∟	_	_	_
		S Paulo	Brasil	10	∟	_	_	_
		Delano	USA	200/250	_		_	_
		Dixon	USA	200				_

MHz	Metres	Station	Country	kW	M	M	S	N
		Riazan	USSR	240	∠	—	—	—
		Novosibirsk	USSR	240/500	—			
		Petropavlo Kam	USSR	240	∠			—
		Poro	Philippines	50	—	—		
		Tirana	Albania					
		S Pedro Sula	Honduras Rep	0.5	/		—	—
		Mahe	Seychelles	100	∠		—	—
		Ankara	Turkey	250	/			—
		Athinai	Greece					
6.190	48.47	S M Galeria	Vatican	100	∠	—	—	—
		Cite Vatican	Vatican	50/80	∠	—	—	—
		Bremen	Germany (W)	10	∠	—	—	—
		Bucuresti	Roumania	250	∠	—	—	—
		Greenville	USA	250/500	∠	—	—	—
		Delhi	India	10	∠	—	—	—
		Omsk	USSR	100	∠	—	—	
		Nikolaevskamur	USSR	50	∠	—		—
		Pt Plata	Dominican Rep	0.1	∠	—	—	—
		Sebaa Aioun	Morocco	10/50	∠	—	—	—
		Osaka	Japan	0.5/6	∠	—	—	—
		Padang	Indonesia	10	∠	—	—	—
		Noblejas	Spain	350	∠	—	—	—
		Tirana	Albania					
		Bonaire Zuid	Neth Antilles	10/100	—			
		Santiago	Chile	25/100	∠	—	—	—
		Kavalla	Greece	250				—
6.195	48.43	London	UK	100/250	∠	—	—	—
		Masirah	Oman	100/200				—
		Kranji	Senegal	50/250	/		—	—
		Antigua	Br W Indies	250	∠	—	—	—
		Limassol	Cyprus	20/100		—	—	—
		Tebrau	Malaysia	100/250		—	—	—
		Baku	USSR	50	∠	—	—	—
		Cali	Colombia	1	∠	—	—	—
		Rio de Janeiro	Brasil	7.5	∠	—	—	—
		La Paz	Bolivia	5	∠	—	—	—
		Sokoto	Niger	10	∠	—	—	—
		Leipzig	Germany (E)	100	—			—
		Nauen	Germany (E)	100	—			—
		Ismaning	Germany (W)	100	∠	—		—
		La Ceiba	Honduras Rep	0.5	/	—		—
		Sackville	Canada	250	∠	—		—
		Warszawa	Poland	1	/	—	—	—
		Pt au Prince	Haiti	50				—
6.200	48.39	Tirane	Albania	240				
		Moskva	USSR					
		Leningrad	USSR					
		Padang	Indonesia					

MHz	Metres	Station	Country	kW
		V Communist Party of Turkey		
		Houa Phan	Laos	
		Bizam Radio		
		Tripoli	Libya	
		Jurong	Singapore	
6.205	48.35		USSR	
		Peking	China Rep	
6.206	48.34	Mebo II		
6.208	48.32	Milan	Italy	
6.210	48.31	Peking	China Rep	
		Vientiane	Laos	
		Vatican	Vatican	
		Paris	France	
		Tirane	Albania	
6.215	48.27		USSR	
		Monte Carlo	Monaco	
		Andorra	Andorra	
6.221	48.22	Wien	Austria	
		Vatican	Vatican	
6.225	48.19	Peking	China Rep	
6.230	48.15	Cairo	Egypt	50
			USSR	
		Vatican	Vatican	
		Kabul	Afghanistan	
6.235	48.12	Karachi	Pakistan	
6.240	48.08	Irkutsk	USSR	
		Seoul	Korea (S)	
6.243	48.05		USSR	
6.250	48.00	Pyongyang	Korea (N)	50
		Malabo	Guinea	10
		V of Peace		
6.260	47.92	Peking	China Rep	
		Cao Lang	Vietnam	
6.270	47.85	Peking	China Rep	
6.280	47.77	Peking	China Rep	
		Nghe Tinh	Vietnam	
6.285	47.73		USSR	
		R Bayrak	Cyprus	
6.290	47.69	Pyongyang	Korea (N)	
		Peking	China Rep	
6.304	47.59	V of People of Burma		
6.320	47.47	Peking	China Rep	
6.330	47.39	Peking	China Rep	
6.331	47.39	Son La	Vietnam	
6.338	47.33	Pyongyang	Korea (N)	
6.340	47.32		Turkey*	
6.345	47.28	Peking	China Rep	

MHz	Metres	Station	Country	kW
6.348	47.26	R Echo of Hope		
6.385	46.99	Ulan Bator	Mongolian Rep	
6.390	46.95		USSR	
6.400	46.88	Pyongyang	Korea (N)	
			USSR*	
6.405	46.84	Taipei	China Nat	
6.410	46.80	Peking	China Rep	
6.426	46.69	Hanoi	Vietnam	
6.430	46.66	Peking	China Rep	
6.435	46.62	Hanoi	Vietnam	
6.453	46.49	German Figure Groups*		
6.495	46.19	Peking	China Rep	
6.520	46.01	Peking	China Rep	
6.540	45.87	Peking	China Rep	
6.548	45.82		USSR	
6.550	45.80	Peking	China Rep	
		Voice of Lebanon		
6.554	45.77		USSR	
6.555	45.77	Peking	China Rep	
6.560	45.73	Met Station*	USSR	
6.568	45.68		USSR	
6.576	45.62	Pyongyang	Korea (N)	
6.585	45.56	Peking	China Rep	
6.590	45.52	Peking	China Rep	
6.596	45.48	Met Station*	USSR	
6.600	45.45	Pyongyang	Korea (N)	
		Pakse	Laos	
6.615	45.35	Met Station*	USSR	
6.617	45.34	Met Station*	USSR	
6.645	45.15	Peking	China Rep	
6.665	45.01	Peking	China Rep	
6.675	44.94	Xieng Khouang	Laos	
6.730	44.58	Met Station*	USSR	
6.745	44.48	Paris*	France	
6.750	44.44	Peking	China Rep	
6.765	44.35	Fukien Front Stn	China Rep	
6.770	44.31		USSR*	
		Pyongyang	Korea (N)	
6.790	44.18	Peking	China Rep	
6.807	44.07	Riyadh*	Saudi Arabia	
6.808	44.07		USSR*	
6.810	44.05	Peking	China Rep	
6.822	43.98		USSR*	
6.825	43.96	Peking	China Rep	
			USSR*	
		V of Democratic		
		Kampuchea		
6.838	43.87	London*	UK	
6.840	43.86	Huhetot	Mongolian Rep	

MHz	Metres	Station	Country	kW
6.852	43.78		USSR*	
		V of Lebanon		
6.860	43.73	Peking	China Rep	
6.865	43.70	Shepparton*	Australia	
6.870	43.67	Alma Ata	USSR	
6.873	43.65	Greenville*	USA	
6.876	43.63	Paris*	France	
6.880	43.60	Peking	China Rep	
6.884	43..59	Bac Thai	Vietnam	
6.890	43.54	Peking	China Rep	
			USSR*	
6.900	43.48	Ankara	Turkey	
6.905	43.45		USSR	
6.910	43.42	Udomsai	Laos	
6.920	43.35		USSR*	
6.935	43.26	Peking	China Rep	
6.937	43.25		China Rep	
6.940	43.23	V of Iraqi Kurdistan		
6.941	43.22	Rhodes	Greece	
6.945	43.20	Moscow	USSR	
6.955	43.13	Peking	China Rep	
6.970	43.04	RFE/R Liberty*		
6.975	43.01	Julich*	Germany (W)	
		RFE/R Liberty*		
		Luang Prabang	Laos	
6.980	42.98		USSR*	
6.987	42.94		USSR*	
6.995	42.89	Peking	China Rep	
		RFE/R Liberty*		
7.010	42.80	Peking	China Rep	
			USSR	
7.015	42.77	V of NUFK		
7.020	42.74	Novosibirsk	USSR	
7.025	42.70	Peking	China Rep	
7.030	42.67		USSR	
		Peking	China Rep	
7.035	42.64	Peking	China Rep	
7.040	42.61	Thimpu	Bhutan	0.3/10
			USSR	
		Peking	China Rep	
		Vatican	Vatican	
7.045	42.58		USSR	
		Peking	China Rep	
7.050	42.55	Cairo	Egypt	100
		Urumchi	China Rep	
7.055	42.52	Peking	China Rep	
		Algiers	Algeria	
7.060	42.49	Peking	China Rep	
7.065	42.46	Tirane	Albania	240

MHz	Metres	Station	Country	kW	M	M	S	N
		Peking	China Rep					
7.075	42.40	Cairo	Egypt	100				
		Tirane	Albania	120				
7.080	42.37	Bac Thai	Vietnam					
		Peking	China Rep	120				
			USSR					
		Tirane	Albania	120				
7.085	42.34	Karachi	Pakistan					
7.090	42.31	Tirane	Albania					
7.095	42.28	Peking	China Rep					
		Karachi	Pakistan					
		Laos	Laos					
7.100	42.25		USSR					
7.103	42.24	V of the People						
7.105	42.22	London	UK	100/250	∠	_	_	_
		Ascension	Ascension	250	∠	_	_	_
		Orcha	Bielorussia USSR	50/240	∠			_
		Simferopol	Ukraine	100/500	/			_
		Brazzaville	Congo	4	∠	_	_	_
		Delhi	India	10/100	_	_	_	_
		Kohima	India	2	/			
		Yogyakarta	Indonesia	20/100	∠	_	_	_
		Arganda	Spain	100	/			
		Noblejas	Spain	350		_		_
		Tananarive	Malagasy Rep	4/30	/	_	_	_
		Ekala	Ceylon	35		_	_	
		Mt Carlo	Monaco	100	∠	_	_	_
		Kathmandu	Nepal	100	∠			
		Julich	Germany (W)	100	∠	_	_	_
		Wertachtal	Germany (W)	500		_	_	
		Quetta	Pakistan	10				_
7.110	42.19	London	UK	100	∠	_	_	_
		Limassol	Cyprus	100	/			_
		Khabarovsk	USSR	100		_	_	
		Omsk	USSR	50/100	∠	_	_	_
		Rhodos	Greece	50	∠	_	_	_
		Warszawa	Poland	20/100	∠	_	_	_
		Bamako	Malawi	50		_	_	
		Ekala	Ceylon	10/35	∠	_	_	_
		Jakarta	Indonesia	50/100	∠	_	_	_
		Riyadh	Saudi Arabia	350	∠		_	
		Maputo	Mozambique	10	_		_	
		Gedja	Ethiopia	100		_	_	
		Arganda	Spain	100	/		_	_
		Delhi	India	10	/			
7.115	42.16	Bangkok	Thailand	5	∠	_	_	_
		Bandundu	Zaire	10	∠	_	_	_
		Biblis	Germany (W)	100	∠	_	_	_

MHz	Metres	Station	Country	kW	M	M	S	N
		Lisbonne	Portugal	100	∠		_	_
		Tbilisi	USSR	240				_
		Duchanbe	USSR	100				_
		Tchita	USSR	100/240	_			
		Ivanofrankovsk	Ukraine	240	∠			_
		Sofia	Bulgaria	500				_
		Gedja	Ethiopia	100			_	_
		Kimjae	Korea (S)	250			_	
7.120	42.13	London	UK	100/250	∠	_	_	_
		Denpassar	Indonesia	10	∠	_	_	_
		N Djamena	Chad	100	∠			_
		Tirane	Albania					
		Bucuresti	Roumania	18	_			
		Mogadiscio	Somalia	50	∠	_	_	_
		Novosibirsk	USSR	100	∠	_		
		Tula	USSR	100	_			
		Lvov	Ukraine	240/500	∠		_	
		Sulaibiyah	Kuwait	250	∠	_		_
		Delhi	India	100	∠	_	_	
		Cyclops	Malta	250	/			
7.125	42.11	Conakry	Guinea	100	∠	_	_	_
		Delhi	India	50/100	∠	_	_	_
		Aligarh	India	250	_	_	_	_
		Nairobi	Kenya	5	∠	_	_	_
		Warszawa	Poland	40/100	∠	_	_	_
		Kazan	USSR	100	_	_		
		Kenga	USSR	100	∠		_	
		Khabarovsk	USSR	240	∠	_	_	
		Jerusalem	Israel	50/300	_			
		Mt Carlo	Monaco	100			_	_
		Athinai	Greece	100	/			
		Kavalla	Greece	250	/			_
7.130	42.08	London	UK	100/250	∠	_	_	_
		Limassol	Cyprus	100	_			
		Stapok	Malaysia	10				
		Wertachtal	Germany (W)	500	∠	_	_	_
		Julich	Germany (W)	100	∠	_	_	_
		Minsk	Bielorussia	100	_	_		
		Erevan	USSR	100	∠		_	
		Vinnitsa	Ukraine	240	_		_	
		Vladivostock	USSR	100			_	
		Sines	Portugal	250	_		_	
		Limbe	Malawi	100	∠		_	_
		Noblejas	Spain	350	∠	_	_	_
		Taipei	China Nat					
		Kavalla	Greece	250	/		_	
			Pakistan					
7.135	42.05	Careysburg	Liberia	250	∠	_	_	_
		Mt Carlo	Monaco	30	∠	_	_	_

MHz	Metres	Station	Country	kW	M	M	S	N
		Islamabad	Pakistan	250	∠		_	_
		Karachi	Pakistan	10	_	_		
		Duchanbe	USSR	240	_	_	_	_
		Moskva	USSR	240				_
		Komsomolskamur	USSR	240	∠			_
		Simferopol	Ukraine	240/500	∠			_
		Tinang	Philippines	250	∠	_	_	_
		Allouis	France	100/500	∠	_	_	_
		Kimjae	Korea (S)	100	∠			
		Delhi	India	100	/			
		Cyclops	Malta	250	/			
		Franceville	Gabon Rep	500			_	
		Thessaloniki	Greece	35	/			
7.140	42.02	London	UK	100/250	_		_	
		Masirah	Oman	100	/		_	_
		Limassol	Cyprus	20/100	∠	_	_	_
		Amboina	Indonesia	10	∠	_	_	_
		Hyderabad	India	10	∠	_	_	_
		Nairobi	Kenya	100	∠	_	_	_
		Kazan	USSR	100		_	_	
		Riga	USSR	240	∠		_	
		Alma Ata	USSR	240				
		Tokyo Yamata	Japan	20/100	∠	_	_	_
		Nampula	Mozambique	5				_
7.145	41.99	Stapok	Malaysia	10				_
		Warszawa	Poland	40/100	∠	_	_	_
		Tachkent	USSR	100		_	_	
		Novosibirsk	USSR	100	/	_		_
		Tula	USSR	100				_
		Quelimane	Mozambique	0.3	∠	_	_	_
		Alger	Algeria	50	∠	_	_	_
		Karlsborg	Sweden	350	_			
		Kavalla	Greece	250	_	_		
		Delhi	India	100	_	_	_	_
		Lisbonne	Portugal	100/250	_		_	_
		Biblis	Germany (W)	100				
		Lampertheim	Germany (W)	100			_	
		Mt Carlo	Monaco	100	_	_		
7.150	41.96	London	UK	100/250	∠	_	_	_
		Ascension	Ascension	125	_			
		Krasnoiarsk	USSR	100	_	_		
		Tchita	USSR	100/500	∠		_	
		Serpukhov	USSR	240	_	_		
		Lvov	Ukraine	240/500	∠		_	_
		Nairobi	Kenya	10	∠	_	_	_
		Julich	Germany (W)	100	/	_	_	
		Wertachtal	Germany (W)	500	∠	_	_	_
		Ismaning	Germany (W)	100				_
		Pemba	Mozambique	0.3	∠	_	_	_

MHz	Metres	Station	Country	kW	M	M	S	N
		Kavalla	Greece	250	_			
		Quetta	Pakistan	10	/			
		Dacca	Bangladesh	10/100	_	_	_	
		Suwon	Korea (S)	50	_	_	_	_
		Bata	Guinea	50	L	_	_	_
		Sofia	Bulgaria	100/500	L	_		_
		Gauhati	India	10	/	_	_	_
7.155	41.93	London	UK	100/250	L	_	_	_
		Amman	Jordan	100	L	_	_	_
		Tananarive	Mongolian Rep	5/30	/	_	_	_
		S M Galeria	Vatican	100	L	_	_	_
		Armavir	USSR	100	L			_
		Poro	Philippines	100	L	_	_	_
		Szekesfehervar	Hungary	20	_	_	_	
		Jaszbereny	Hungary	250	L	_	_	
		Arganda	Spain	100	L	_	_	_
		Playa de Pals	Spain	250/500				_
		Biblis	Germany (W)	100			_	
		Holzkirchen	Germany (W)	10			_	
		Lampertheim	Germany (W)	100	L	_	_	
		Ismaning	Germany (W)	100			_	
7.160	41.90	Stapok	Malaysia	10			_	
		Madras	India	10	_	_		
		Kazan	USSR	240	/	_	_	_
		Omsk	USSR	100	_	_	_	
		Tula	USSR	240	/	_	_	_
		Petropavlo Kam	USSR	100				_
		Lvov	Ukraine	240	_			
		Allouis	France	100/500	L	_	_	_
		Hargeisa	Somalia	10	L	_	_	_
		Tinang	Philippines	250	_	_		_
		Kavalla	Greece	250	L	_	_	_
		Julich	Germany (W)	100	L	_	_	_
		Cyclops	Malta	250	L	_	_	_
		S M Galeria	Vatican	100	L	_	_	_
		Huambo	Angola	10	/		_	_
7.165	41.87	Tripoli	Libya	100	L	_	_	
		Dar es Salaam	Tanzania	20	_	_	_	
		Delhi	India	20/100	_	_	_	_
		Jayapura	Indonesia	1/5	L	_	_	
		Lampertheim	Germany (W)	100	/			
		Biblis	Germany (W)	100	/	_	_	
		Holzkirchen	Germany (W)	10	L	_	_	_
		Playa de Pals	Spain	250			_	
		Lisbonne	Portugal	100/250	L	_	_	
		Kiev	Ukraine	240	L	_	_	_
		Serpukhov	USSR	240				_
		Gedja	Ethiopia	100	/	_	_	_
		Islamabad	Pakistan	100	_	_	_	_

MHz	Metres	Station	Country	kW	M	M	S	N
		Karachi	Pakistan	50	—	—		—
		Poro	Philippines	50	—	—	—	
		Kathmandu	Nepal	100	∠		—	—
7.170	41.84	London	UK	250	—		—	
		Kavalla	Greece	250	∠	—	—	
		Dakar	Senegal	30	∠	—	—	—
		Noumea	New Caledonia	20	∠	—	—	—
		Singapore	Singapore	10	∠	—	—	—
		Armavir	USSR	240	—		—	
		Kazan	USSR	100	—	—		
		Novosibirsk	USSR	100	∠			
		Tachkent	USSR			—		—
		Meyerton	South Africa	100	—	—		—
		Tanger	Morocco	100				—
		Wien	Austria	100	∠	—	—	—
		Ankara	Turkey	250	/	—	—	—
		Ranchi	India	2	/			
7.175	41.81	Brazzaville	Congo	4/25	∠	—	—	—
		Caltanissetta	Italy	5	∠	—	—	—
		Wertachtal	Germany (W)	500	∠	—	—	—
		Starobelsk	Ukraine	240/500	∠	—		—
		Gorkii	USSR	240	∠	—	—	—
		Khabarovsk	USSR	100	∠	—	—	—
		Kinghisepp	USSR	240	/			—
		Careysburg	Liberia	250	∠	—	—	—
		Bucuresti	Roumania	18/120	∠	—	—	—
		Gwelo	Rhodesia	10/100	/		—	
7.180	41.78	London	UK	100/250	—			
		Kranji	Singapore	250	∠	—	—	—
		Warszawa	Poland	100	∠	—	—	—
		Abu Ghraib	Iraq	100	—	—	—	
		Babel	Iraq	500	—	—		
		Bhopal	India	10	∠	—	—	—
		Kazan	USSR	100	—			
		Tanger	Mongolian Rep	100		—		—
		Gedja	Ethiopia	100	∠			
		Biblis	Germany (W)	100	—			—
		Lampertheim	Germany (W)	100	/		—	
		Kavalla	Greece	250				—
		V of the Namibian People						
7.185	41.75	London	UK	100/250	∠	—	—	—
		K Wusterhausen	Germany (E)	5	∠	—	—	—
		Merauke	Indonesia	1	∠	—	—	—
		Gorkii	USSR	100	/			—
		Tbilisi	USSR	100	—		—	—
		Alma Ata	USSR	100		—		—
		Novosibirsk	USSR	240/500	∠	—	—	—
		Rangoon	Burma	50	∠	—	—	—

MHz	Metres	Station	Country	kW	M	M	S	N
		Noblejas	Spain	350	_			
7.190	41.72	Limassol	Cyprus	100		_		
		Ekala	Ceylon	10	/	_	_	_
		Jayapura	Indonesia	10	/	_	_	_
		Armavir	USSR	100	_			
		Moskva	USSR	240		_	_	
		Peking	USSR					
		Tanger	Morocco	100	/	_	_	_
		Aden	Aden	100				
		Bata	Guinea	50	/	_	_	_
		Lisbonne	Portugal	100/250	/	_	_	_
		Holzkirchen	Germany (W)	10			_	
		Lampertheim	Germany (W)	100				_
		Biblis	Germany (W)	100	/	_	_	
		Playa de Pals	Spain	500				
		Poro	Philippines	50	/			_
7.195	41.70	Bucuresti	Roumania	18/250	/	_	_	_
		Delhi	India	100	/	_	_	_
		Careysburg	Liberia	50/250	/	_	_	_
		Tula	USSR	100	/		_	_
		Simferopol	Ukraine	500				_
		Alger	Algeria	50	/	_	_	_
		Noblejas	Spain	350	/	_	_	_
		Tokyo Yamata	Japan	20	/	_	_	_
		Islamabad	Pakistan	100		_	_	
		Karachi	Pakistan	10	/			
7.200	41.67	London	UK	250	/	_	_	_
		Penang	Malaysia	10	/	_	_	_
		Irkutsk	USSR	50	/	_	_	_
		Kazan	USSR	240				_
		Jigulevsk	USSR	100		_	_	
		Vladivostock	USSR	50	/	_	_	_
		Belgrade	Yugoslavia	10	/	_	_	_
		Omdurman	Sudan	120		_	_	
		Kabul	Afghanistan	10	/	_	_	_
		Holzkirchen	Germany (W)	10				_
		Lampertheim	Germany (E)	100	_			
		Lisbonne	Portugal	250				_
		Diosd	Hungary	100	/	_	_	_
		Arganda	Spain	100	_	_	_	_
		Silinhot	Mongolian Rep					
		Tripoli	Libya					
		Jerusalem	Israel					
		V of Blackman's Resistance						
			Philippines					
7.205	41.64	London	UK	250		_	_	_
		Armavir	USSR	100/500	/	_	_	_
		Frunze	USSR	100				_

84

MHz	Metres	Station	Country	kW	M	M	S	N
		Moskva	USSR	240	∟	_	_	
		Athinai	Greece	100	∟	_	_	_
		Kavalla	Greece	250	/	_	_	_
		Rhodos	Greece	50	∟	_	_	_
		Warszawa	Greece	1/100	/	_	_	_
		Yaounde	Cameroon	30	∟	_	_	_
		Lubumbashi	Zaire	10	∟	_	_	_
		Tokyo Yamata	Japan	20	∟	_	_	_
		Ismaning	Germany (W)	100	/	_		
		Kamalabad	Iran	100	_			
7.210	41.61	London	UK	100/250	∟	_	_	_
		Limassol	Cyprus	100	/	_	_	_
		Beromunster	Switzerland	150	∟	_	_	
		Schwarzenburg	Switzerland	150			_	
		Calcutta	India	10	∟	_	_	_
		Moskva	USSR	100	∟	_	_	
		Kazan	USSR	240		_	_	
		Khabarovsk	USSR	50	∟	_	_	_
		Sverdlovsk	USSR	100	_			
		Nairobi	Kenya	10	∟	_	_	_
		Biak	Indonesia	1	∟	_	_	_
		Lopik	Netherlands	100	∟	_	_	_
		Wertachtal	Germany (W)	500	_			
		Julich	Germany (W)	100	∟	_	_	_
		Dakar	Senegal	100	∟	_	_	_
		Ziguinchor	Senegal	30	/			
		Tanger	Morocco	100	_		_	_
		Karachi	Pakistan	50	_			
7.215	41.58	Abidjan	Ivory Coast	10	∟	_	_	_
		Delhi	India	100/250	_	_	_	_
		Biblis	Germany (W)	100	/			
		Lisbonne	Portugal	50/100	_		_	
		Jaszbereny	Hungary	250	∟	_	_	_
		Vinnitsa	Ukraine	240/500	_			
		Armavir	USSR	240				
		Komsomolskamur	USSR	100		_	_	
		Athinai	Greece	100		_		
		Kavalla	Greece	100/250	_	_	_	_
		Kamalabad	Iran	100	_	_	_	_
		Berakas	Brunei	10	_	_	_	
		Sofia	Bulgaria	50	_			
		Manzini	Swaziland	25	_	_	_	
		Mt Carlo	Monaco					
7.220	41.55	London	UK	250	/	_		
		Lusaka	Zambia	20	∟	_	_	_
		Diriyya	Saudi Arabia	50	∟	_	_	_
		Bangui	Central African Rep	4/100		_	_	_

MHz	Metres	Station	Country	kW	M	M	S	N
		Tanger	Morocco	50/100	_		_	_
		Tachkent	USSR	100				_
		Tchita	USSR	240/500	∟	_	_	_
		Vladivostock	USSR	100	_			
		Jakarta	Indonesia	1	∟	_	_	_
		Biblis	Germany (W)	100				
		Lampertheim	Germany (W)	100	∟	_	_	_
		Holzkirchen	Germany (W)	10	/		_	
		Playa de Pals	Spain	250/500	∟			_
		Salman Pak	Iraq	100	_			
7.225	41.52	Bucuresti	Roumania	120/250	∟	_	_	_
		Delhi	India	100	_	_	_	_
		Aligarh	India	250	∟	_	_	_
		Bocaue	Philippines	25/50	∟	_	_	_
		Sebaa Aioun	Morocco	10	∟	_	_	_
		Kigali	Rwanda	250	∟	_	_	_
		Abu Zabaal	Egypt	100	_			
		Male	Maldives	2.7	_	_	_	_
		Arganda	Spain	100	_	_	_	_
		Sfax	Tunisia					
7.230	41.49	London	UK	100/250	∟	_	_	_
		Masirah	Oman	100				
		Limassol	Cyprus	20/100	∟	_	_	_
		Kazan	USSR	100	∟		_	_
		Frunze	USSR	100	_			
		Krasnoiarsk	USSR	50	∟	_	_	_
		Nikolaevskamur	USSR	50/100	_	_	_	
		Kiev	Ukraine	50	_			
		Lvov	Ukraine	240		_		
		Mt Carlo	Monaco	100	∟	_	_	_
		Tananarive	Malagasy Rep	10	/			_
		Ouagadougou	Upper Volta	20	_			
		Tanger	Morocco	50/100	∟	_	_	_
		Dar es Salaam	Tanzania	50	_	_	_	_
		Kamalabad	Iran	100	_		_	_
		Kurseong	India	20	/			
7.235	41.47	London	UK	100	∟	_	_	_
		Julich	Germany (W)	100	∟	_	_	_
		Wertachtal	Germany (W)	500	∟	_	.	_
		Delhi	India	50	_	_	_	_
		Roma	Italy	100	∟	_	_	_
		Enugu	Niger	10	∟	_	_	_
		S M Galeria	Vatican	100	∟	_	_	_
		Lusaka	Zambia	50	∟	_	_	_
		Riga	USSR	100	_	_		
		Sverdlovsk	USSR	100	∟			
		Poro	Philippines	35/50	/	_		_
		Luanda	Angola	100	/		_	_
		R Kulmis						

MHz	Metres	Station	Country	kW	M	M	S	N
7.240	41.44	London	UK	150	/		_	_
		Garoua	Cameroon	4	∠	_	_	_
		Maputo	Mozambique	25/100	∠	_	_	_
		Belgrade	Yugoslavia	10	∠	_	_	_
		Bombay	India	10	∠	_	_	_
		Medan	Indonesia	7.5	∠	_	_	_
		Tula	USSR	240/500	∠	_	_	_
		Nairobi	Kenya	10	∠	_	_	_
		Kavalla	Greece	250	∠		_	
		Carnarvon	Australia	100	_	_	_	_
		Shepparton	Australia	10	_	_	_	_
		Lyndhurst	Australia					
		Tanger	Morocco	100				_
		Poro	Philippines	50	_	_	_	_
		Lopik	Netherlands	100	∠	_	_	_
		Kimjae	Korea (S)	250	∠	_	_	_
		V of Eritrea Revln						
		Salman Pack	Iraq	100		_	_	
7.245	41.41	London	UK	100	_	_		
		Nouakchott	Mauretania	4/100				_
		Arganda	Spain	100	/		_	
		Playa de Pals	Spain	250	_		_	
		Holzkirchen	Germany (W)	10	∠	_	_	_
		Biblis	Germany (W)	100	/	_	_	_
		Lampertheim	Germany (W)	100	∠		_	
		Lisbonne	Portugal	50			_	
		Luanda	Angola	100			_	
		S Denis	Reunion	4	∠	_	_	_
		Vinnitsa	Ukraine	240	_			
		Khabarovsk	USSR	240		_	_	
		Krasnoiarsk	USSR	50				_
		Mt Carlo	Monaco	100	∠	_		_
		Alger	Algeria	50	∠	_		_
		Praha	Czechoslovakia	120	∠		_	
		Kimjae	Korea (S)	100		_		
		Dacca	Bangladesh	10	/		_	
7.250	41.38	London	UK	100	_	_	_	_
		Limassol	Cyprus	20/100	/		_	
		Masirah	Oman	100	∠	_	_	_
		Singapore	Singapore	50	∠	_	_	_
		Sverdlovsk	USSR	100				_
		Tula	USSR	240/500		_	_	
		S M Galeria	Vatican	100	∠	_	_	_
		Lusaka	Zambia	120	∠	_	_	_
		Tokyo Yamata	Japan	100		_		
		Kimjae	Korea (S)	100	∠	_	_	
		Lucknow	India	10	∠	_	_	_
7.255	41.35	London	UK	100	∠	_	_	_
		Nampula	Mozambique	0.3	∠	_	_	_

MHz	Metres	Station	Country	kW	M	M	S	N
		Ikorodu	Niger	100	/		_	_
		Sogunle	Niger	10		_		_
		Sofia	Bulgaria	50	∠	_		
		Biblis	Germany (W)	100	∠	_	_	_
		Lampertheim	Germany (W)	100				
		Holzkirchen	Germany (W)	10		_	_	
		Lisbonne	Portugal	100	∠		_	_
		Kenga	USSR	50		_		
		Tbilisi	USSR	50	∠			_
		Manzini	Swaziland	25		_		
		Kinshasa	Zaire	10	∠	_	_	
		Poro	Philippines	50/100		_	_	
7.260	41.32	London	UK	100/250	∠	_	_	_
		Limassol	Cyprus	20/100	∠	_	_	_
		Mt Carlo	Monaco	100	∠	_	_	_
		Port Vila	New Hebrides	2		_	_	_
		Bombay	India	100	/			
		Delhi	India	10/50		_	_	_
		Madras	India	100		_	_	_
		Iujnsakhalinsk	USSR	100	_			_
		Minsk	Bielorussia	100	∠	_	_	
		Ulan Bator	Mongolian Rep	25				
		Nauen	Germany (E)	500	∠	_	_	_
7.265	41.29	London	UK	250				_
		Togblekope	Togo	100	∠		_	_
		Riazan	USSR	240	_		_	
		Iakutsk	USSR	100	∠	_	_	_
		Komsomolskamur	USSR	240		_	_	
		Armavir	USSR	240	_	_	_	
		Volgograd	USSR	100		_		
		Rohrdorf	Germany (W)	20	∠	_	_	
		Karachi	Pakistan	10/50		_	_	_
		Islamabad	Pakistan	100/240	/			
		Tanger	Morocco	50				
		Cyclops	Malta	250	∠	_	_	
		Sanaa	Yemen	50		_		
		Luanda	Angola		/			
7.270	41.27	Ascension	Ascension	250		_	_	_
		Limassol	Cyprus	100		_		
		Stapok	Malaysia	10		_		
		Meyerton	South Africa	20/500	∠	_	_	_
		Jakarta	Indonesia	50/100	∠	_	_	_
		Erevan	USSR	100	_			_
		Kenga	USSR	100/500	∠		_	
		Moskva	USSR	50				
		Kavalla	Greece	250	∠	_	_	_
		Warszawa	Poland	100	∠	_	_	_
		Sofia	Bulgaria	50/100		_		
		Ankara	Turkey	250		_	_	_

MHz	Metres	Station	Country	kW	M	M	S	N
		Kamalabad	Iran	100	–			
		Delhi	India	100				
		Franceville	Gabon	500			–	–
		Ismaning	Germany (W)					
7.275	41.24	Masirah	Oman	200	/	–	–	–
		Limassol	Cyprus	100	–	–		–
		Ikorodu	Niger	100	–	–		
		Tinang	Philippines					–
		Poro	Philippines	50	–	–	–	
		Roma	Italy	60/100	∠	–	–	–
		Kenga	USSR	100	–	–		
		Komsomolskamur	USSR	240	/			
		Krasnoiarsk	USSR	100				
		Duchanbe	USSR	50	∠	–	–	–
		Ismaning	Germany (W)	100				–
		Wertachtal	Germany (W)	500	∠	–	–	–
		Julich	Germany (W)	100	∠	–	–	
		Tirane	Albania					
		Cyclops	Malta	250	∠	–	–	–
		Mt Carlo	Monaco	100				–
		Sfax	Tunisia	100			–	–
		Jaszbereny	Hungary	250	∠			–
		Arganda	Spain	100	∠	–	–	–
		Tanger	Morocco	100		–	–	
		Kavalla	Greece	250	∠	–	–	–
		Suwon	Korea (S)	50	/			
7.280	41.21	London	UK	250	–			
		Moskva	USSR	200	∠	–	–	–
		Komsomolskamur	USSR	240		–	–	
		Dar es Salaam	Tanzania	10		–	–	–
		Careysburg	Liberia	250	∠	–	–	–
		Tirane	Albania					
		Thessaloniki	Greece	35	–	–	–	–
		Manzini	Swaziland	25	–			–
		Gauhati	India	10	/			
		Delhi	India	20	∠	–	–	
		Aligarh	India	250	∠	–	–	–
		Allouis	France	100/500	∠	–	–	–
		Dacca	Bangladesh	100	/			
7.285	41.18	London	UK	100	∠	–	–	–
		Warszawa	Poland	15	∠	–	–	–
		Tula	USSR	100		–		
		Kazan	USSR	100	–			
		Krasnoiarsk	USSR	100	/			–
		Moskva	USSR			–		
		Allouis	France	100	–	–	–	–
		Ibadan	Niger	1/10	∠	–	–	–
		Wertachtal	Germany (W)	500	∠	–	–	–
		Julich	Germany (W)	100	–	–	–	

MHz	Metres	Station	Country	kW	M	M	S	N
		Sines	Portugal	250	L	_	_	_
		Talata Volon	Malagasy Rep	600	L	_	_	_
		Bamako	Malawi	18		_	_	
		Gwelo	Rhodesia	20/100	/		_	
		Poro	Philippines	100				_
7.290	41.15	Moskva	USSR	240/500	L	_	_	_
		Petropavlo Kam	USSR	240	L			
		Delhi	India	60/100	L	_	_	_
		Roma	Italy	60/100	L	_	_	_
		Tirane	Albania					
		Ismaning	Germany (W)	100/250	L	_		
		Islamabad	Pakistan	100/250	L	_	_	_
		Kimjae	Korea (S)	100	L	_	_	_
		Dacca	Bangladesh	100	/			/
7.295	41.12	London	UK	100/250	L	_	_	_
		Accra	Ghana	10	L	_	_	_
		K Wusterhausen	Germany (E)	100	L	_	_	_
		Kajang	Malaysia	100	L	_	_	_
		Moskva	USSR	100/240	L	_	_	_
		Blagovechtchen	USSR	100		_	_	
		Menado	Indonesia	0.5	L	_	_	_
		Mbujimayi	Zaire	10	L	_	_	_
		Nairobi	Kenya	5	L	_	_	_
		Sines	Portugal	250		_		
		Tanger	Morocco	35	/	_		_
		Biblis	Germany (W)	100	L		_	_
		Holzkirchen	Germany (W)	10	L			
		Lampertheim	Germany (W)	50/100		_	_	_
		Playa de Pals	Spain	250				_
		Lisbonne	Portugal	100		_		
7.300	41.10	Tirane	Albania	240				
		Berlin	Germany (E)					
		Kiev	USSR					
		Moskva	USSR					
7.305	41.07		USSR					
		Leningrad	USSR					
		V of Malayan Revln						
7.310	41.04	Kalinin	USSR					
			USSR					
7.315	41.01	Peking	China Rep					
		Chita	USSR					
7.320	40.98	London	UK					
		Magadan	USSR					
		Vladivostock	USSR					
7.325	40.96	London	UK					
		Moskva	USSR					
		Lanchow	China Rep					

MHz	Metres	Station	Country	kW
7.330	40.93	Minsk/Duchanbe	USSR	50
		Peking	China Rep	
7.335	40.90		USSR	
		Ottawa*	Canada	
7.340	40.87	Moskva	USSR	
7.345	40.84	Praha	Czechoslovakia	100
			USSR	
7.350	40.82	Moskva	USSR	
		V of Kampuchean People		
7.355	40.79	Moskva	USSR	
7.360	40.76	Peking	China Rep	
		Moskva/Kiev	USSR	
7.365	40.73	Rabat*	Morocco	
7.370	40.71	Minsk	USSR	100
			USSR	
7.375	40.68	Hanoi	Vietnam	
		Karachi	Pakistan	
		Peking	China Rep	
7.380	40.65	Moskva	USSR	100
		Peking	China Rep	
7.385	40.62	Peking	China Rep	
		Savannakhet	Laos	
		Hanoi	Vietnam	
			USSR	
7.390	40.60	Kiev	Ukraine	240
7.395	40.57	Moskva	USSR	
		Jerusalem	Israel	
7.400	40.54	Moskva	USSR	
7.410	40.49	Moskva*	USSR	
7.412	40.48	Jerusalem	Israel	
		Delhi	India	100
7.415	40.46	Hanoi	Vietnam	
7.420	40.43	Minsk	USSR	
		Moskva	USSR	100
7.430	40.38	Moskva	USSR	
7.437	40.34		USSR*	
7.440	40.32	Peking	China Rep	
		Moskva	USSR	100
		R.F.E.		
7.442	40.31	Monrovia	Liberia	
7.443	40.31	Geneva (UNO)	Switzerland	25
7.465	40.19	Jerusalem	Israel	
7.470	40.16	Peking	China Rep	
		Hanoi	Vietnam	
7.480	40.11	Peking	China	
		Vientiane	Laos	
7.490	40.05	Vladivostock*	USSR	
7.500	40.00	Lyndhurst Time Sig	Australia	

MHz	Metres	Station	Country	kW
7.504	39.98	Peking	China Rep	
7.512	39.94	Hanoi	Vietnam	
7.530	39.84		USSR*	
7.540	39.79		USSR*	
7.550	39.74		USSR*	
		Seoul	Korea (S)	
		Peking	China Rep	
7.588	39.54	R Sandino		
7.590	39.53	Peking	China Rep	120
7.605	39.45		USSR	
7.620	39.37	Peking	China Rep	
7.649	39.22	Hakkari	Turkey	1
7.651	39.21	Greenville*	USA	
7.660	39.16	Peking	China Rep	
7.670	39.11	Sofia	Bulgaria	
			USSR	
7.675	39.09	Julich	Germany (W)	
7.700	38.96	Peking	China Rep	
7.727	38.82	Ismaning*	Germany (W)	
7.740	38.76		USSR	
7.767	38.62	Julich*	Germany (W)	
7.768	38.62	Greenville*	USA	
7.770	38.61	Peking	China Rep	
		Greenville*	USA	
7.775	38.59	Peking	China Rep	
7.780	38.56	Peking	China Rep	
7.800	38.46	Peking	China Rep	
7.815	38.39	Peking	China Rep	
7.820	39.36	Peking	China Rep	
7.827	38.33	Peking	China Rep	
7.844	38.25	London*	UK	
7.848	38.23	London*	UK	
7.850	38.22	Fukien Front Stn	China Rep	
7.855	38.19	Peking	China Rep	
7.893	38.01	London*	UK	
7.925	37.85		USSR*	
7.935	37.81	Peking	China Rep	
7.948	37.75		USSR*	
7.973	37.63	London*	UK	
7.976	37.61	London*	UK	
7.991	37.54	London*	UK	
8.007	37.47	Peking	China Rep	
8.063	37.21	Algiers	Algeria	
8.125	36.92		USSR*	
8.240	36.41	Peking	China Rep	
8.260	36.32	Peking	China Rep	
8.300	36.14	Peking	China Rep	
8.320	36.06	Peking	China Rep	
8.345	35.95	Peking	China Rep	

MHz	Metres	Station	Country	kW
8.360	35.89	V of Arabian Peninsula People		
8.395	35.74	Luang Prabang	Laos	
8.425	35.61	Peking	China Rep	
8.450	35.50	Peking*	China Rep	
8.490	35.34	Peking	China Rep	
8.565	35.03	Peking	China Rep	
8.600	34.88	Peking	China Rep	
8.660	34.64	Peking*	China Rep	
			Laos	
8 903	33.70		USSR	
8.910	33.67		USSR	
8.917	33.64		USSR	
8.965	33.46			
8.970	33.44		USSR	
9.009	33.30	Jerusalem	Israel	50/100
9.020	33.26	Peking	China Rep	
9.022	33.25	Teheran	Iran	350
9.030	33.22	Peking	China Rep	
9.064	33.10	Peking	China Rep	
9.080	33.04	Peking	China Rep	
9.090	33.00	RFE*		
9.091	33.00	RFE*		
9.130	32.86	Moskva*	USSR	
9.150	32.79	Moskva*	USSR	
9.170	32.72	Peking	China Rep	
		RFE/R Liberty*		
9.200	32.61		USSR*	
9.210	32.57		USSR*	
9.240	32.47		USSR*	
9.250	32.43	RFE/R Liberty*		
9.290	32.29	Peking	China Rep	
9.317	32.20	London*	UK	
9.323	32.17	London*	UK	
9.330	32.15			
9.336	32.13	Peking	China Rep	
9.340	32.12	Peking	China Rep	
			USSR	
9.345	32.10		USSR	
9.355	32.07	Jerusalem	Israel	
9.360	32.05	Madrid	Spain	50
9.365	32.03	Peking	China Rep	
9.375	32.00	Tirane	Albania	
9.380	31.98	Peking	China Rep	
9.390	31.95	Peking	China Rep	
			USSR	
9.400	31.91	Peking	China Rep	
9.410	31.88	London	UK	
9.417	31.86	Peking	China Rep	

MHz	Metres	Station	Country	kW	M	M	S	N
9.420	31.85	Pyongyang	Korea (N)					
		V of People of Thailand						
9.425	31.83	Jerusalem	Israel					
9.430	31.81	Tirane	Albania					
9.435	31.80	Jerusalem	Israel					
9.440	31.78	Peking	China Rep					
			USSR					
9.445	31.76	Jerusalem	Israel					
9.450	31.75	Moskva	USSR	100				
9.455	31.73	Cairo	Egypt	10				
		Peking	China Rep					
9.460	31.71	Peking	China Rep					
			USSR					
		Karachi	Pakistan					
9.465	31.70	Karachi	Pakistan					
9.470	31.68	Peking	China Rep					
		V of Democratic Kampuchea						
9.475	31.66	Cairo	Egypt					
9.480	31.65	Peking	China Rep	120				
		Moskva	USSR	120				
		Tirane	Albania					
9.485	31.63	Tirane	Albania					
9.490	31.61	Lhasa	Tibet					
		Moskva	USSR					
		Peking	China Rep					
9.495	31.60	Cairo	Egypt	100				
		Jerusalem	Israel					
9.500	31.58	Berlin	Germany (E)	100				
		Moskva	USSR	100				
		Bizam Radio		15				
		Tirane	Albania	240				
		Budapest	Hungary	100/250	_		_	
		Tripoli	Libya					
		Dacca	Bangladesh					
		R of the Patriots						
9.505	31.56	London	UK	100	_			
		La Paz	Bolivia	5/100	L	_	_	_
		Bocaue	Philippines	50	L	_	_	_
		Belgrade	Yugoslavia	10	L	_	_	_
		K Wusterhausen	Germany (E)	100	L	_	_	_
		Omdurman	Sudan	50	L	_	_	
		Velkekostolany	Czechoslovakia	120	L	_	_	
		S Domingo	Dominican Rep	20	L	_	_	_
		Tokyo Yamata	Japan	100	L	_	_	_
		S Paulo	Brasil	7.5	L	_	_	_
		Alma Ata	USSR	100	L	_	_	_
		Komsomolskamur	USSR	240	L			_

MHz	Metres	Station	Country	kW	M	M	S	N
		Kenga	USSR	500	/			
		Allouis	France	100/500			_	_
		Noumea	New Caledonia	4	∟		_	_
		Arganda	Spain	100	∟	_	_	_
		Noblejas	Spain	350		_	_	_
		Redwood City	USA	250		_		
		Shepparton	Australia	100	∟	_		_
		Lisbonne	Portugal	50/250	∟		_	_
		Biblis	Germany (W)	100	∟	_	_	_
		Holzkirchen	Germany (W)	10	∟	_		
		Lampertheim	Germany (W)	100				_
		Sulaibiyah	Kuwait	250	∟	_	_	_
		Islamabad	Pakistan	100	/			
9.510	31.55	London	UK	100/250	∟	_	_	_
		Antigua	Br W Indies	250	∟	_	_	_
		Sackville	Canada	50/250	∟	_	_	
		Alger	Algeria	50	∟	_	_	_
		Bucuresti	Roumania	18/250	∟	_	_	_
		Madras	India	100	∟	_	_	_
		Serpukhov	USSR	100	/			
		Vladivostock	USSR	100				
		Irkutsk	USSR	240/500			_	_
		Kazan	USSR	100	∟	_	_	_
		Greenville	USA	50/250	_	_	_	_
		Tanger	Morocco	100	_			
		K Wusterhausen	Germany (E)	100	_			_
		Taipei	China Nat					
		Noblejas	Spain	350	_		_	_
		Santiago	Chile	100	∟	_		_
		Allouis	France	100/500	∟	_	_	_
		V of Arab	Syria					
		V of the One	Lebanon					
		Kimjae	Korea (S)	100				_
		Mt Carlo	Monaco	100	/		_	_
		Wertachtal	Germany	100			_	_
		Padang	Indonesia					
9.515	31.53	London	UK	250				_
		Kajang	Malaysia	50	∟	_	_	_
		Ankara	Turkey	100/250	/	_	_	_
		Caltanissetta	Italy	25	∟	_	_	_
		Erevan	USSR	100			_	_
		Simferopol	Ukraine	100	_	_		
		Mexico	Mexico	20	∟	_	_	_
		Montevideo	Uruguay	10	∟	_	_	_
		Rio de Janeiro	Brasil	1/7.5	∟	_	_	_
		Kampala	Uganda	250				_
		Las Mesas	Canary Is	50			_	_
		Malolos	Philippines	100	_	_	_	_
		Tirana	Albania					

MHz	Metres	Station	Country	kW	M	M	S	N
		Peking	China Rep					
		Meyerton	South Africa	100				_
		Athinai	Greece	100	/			_
		Madras	India	100	/			
9.520	31.51	Pt Moresby	Papua New Guinea	10/50	∟	_	_	_
		Bonanza	Nicaragua	0.1	∟	_	_	_
		Manzini	Swaziland	25		_		
		Abu Zaabal	Egypt	100				
		Blagovechtchen	USSR	100		_	_	
		Kazan	USSR	100				_
		Armavir	USSR	150/500	∟	_	_	_
		Tachkent	USSR	100	_			
		Holzkirchen	Germany (W)	10		_		
		Lampertheim	Germany (W)	50/100	/	_	_	_
		Biblis	Germany (W)	100		_	_	_
		Playa de Pals	Spain	250/500	_	_	_	_
		Noblejas	Spain	350	∟	_	_	_
		Santiago	Chile	100	/			_
9.525	31.50	Scituate	USA	50/100	_			
		Okeechobee	USA	100				_
		Cincinnati	USA	250	∟	_	_	_
		Habana	Cuba	50	∟	_	_	_
		Tokyo Yamata	Japan	100	∟	_	_	_
		Warszawa	Poland	100	∟	_	_	_
		Aligarh	India	250	_	_	_	_
		Bombay	India	100	∟	_	_	_
		Delhi	India	100	/			
		Mt Carlo	Monaco	100	_			_
		Allouis	France	100/500	_	_	_	_
		Las Mesas	Canary Is	50	_	_	_	_
		Kimjae	Korea (S)	180/250	∟	_	_	_
		Peking	China Rep					
9.530	31.48	London	UK	100/250				_
		Limassol	Cyprus	100				_
		Amman	Jordan	100	∟	_	_	_
		Greenville	USA	50/500	∟	_	_	_
		Redwood City	USA	250	_			
		Dar es Salaam	Tanzania	20	_	_	_	_
		Vinnitsa	Ukraine	50	_			
		Moskva	USSR	240/500	∟	_	_	_
		Vladivostock	USSR	100	∟			
		Komsomolskamur	USSR	100		_		
		Frunze	USSR	100/500				_
		Poro	Philippines	50	_	_	_	_
		Tokyo Yamata	Japan	100	∟	_	_	_
		Tanger	Morocco	35/100		_		
		Bucuresti	Roumania	18/120	∟	_	_	_

MHz	Metres	Station	Country	kW	M	M	S	N
		Santiago	Chile	10/100	_			
		Athinai	Greece	100	∠	_	_	_
		Kavalla	Greece	250				_
		Peking	China Rep					
		Sackville	Canada	50/250	_	_		_
		Sofia	Bulgaria	500	_	_		_
		Noblejas	Spain	350	∠	_	_	_
		Singapore	Singapore	50	∠	_	_	_
		Cyclops	Malta	250			_	
		Dacca	Bangladesh	100	/			_
		Franceville	Gabon Rep	500			_	
		Ismaning	Germany (W)	100				_
		Calcutta	India	10	/			
9.535	31.46	Stapok	Malaysia	10			_	
		Schwarzenburg	Switzerland	150	_			
		Sarnen	Switzerland	250	∠	_	_	_
		Aligarh	India	250	/			_
		Delhi	India	50/100	_	_	_	_
		Luanda	Angola	100	/		_	_
		Nagoya	Japan	0.6	∠	_	_	_
		Simferopol	Ukraine	100	∠			_
		Sackville	Canada	50/250	∠			_
		Bonaire Zuid	Neth Antilles	50/250			_	_
		K Wusterhausen	Germany (E)	100				
		Allouis	France	100/500	_	_	_	
		Kimjae	Korea (S)	100	∠	_	_	_
		Kamalabad	Iran	100	_			
		Malolos	Philippines	100	∠	_	_	_
		Dacca	Bangladesh	100	/			
		Peking	China Rep					
9.540	31.45	London	UK	250				_
		Limassol	Cyprus	100	_	_	_	_
		Lyndhurst	Australia	10	∠	_	_	_
		Minsk	Bielorussia	50	/	_	_	
		Moskva	USSR	50/240	∠	_	_	
		Tachkent	USSR	100	∠	_	_	_
		Petropavlo Kam	USSR	200	∠	_	_	_
		Warszawa	Poland	8/100	∠	_	_	_
		Velkekostolany	Czechoslovakia	120	∠	_	_	_
		Tanger	Morocco	25/100				_
		Kavalla	Greece	250	∠	_	_	
		Allouis	France	100/500	_	_	_	_
		Ismaning	Germany (W)	100	_	_		
		Lampertheim	Germany (W)	100				_
		Malolos	Philippines	100	∠	_	_	_
		Greenville	USA	50	∠	_	_	_
		Bucuresti	Roumania	120/250	∠			_
		Kimjae	Korea (S)	100			_	

MHz	Metres	Station	Country	kW	M	M	S	N
		Jerusalem	Israel	20/300				—
		Peking	China Rep					
		Careysburg	Liberia	250	—			
		Dacca	Bangladesh	100	/			—
9.545	31.43	Antigua	Br W Indies	250	∠	—	—	—
		Montserrat	Br W Indies	15/50	∠	—	—	—
		Khabarovsk	USSR	100			—	—
		Tula	USSR	50	/			—
		Tema	Ghana	100	∠	—	—	—
		Curityba	Brasil	7.5	∠	—	—	—
		Julich	Germany (W)	100	∠	—	—	—
		Wertachtal	Germany (W)	500	∠	—	—	—
		Delano	USA	250	∠	—	—	—
		Dixon	USA	200				—
		Vera Cruz	Mexico	0.5	∠	—	—	—
		Tinang	Philippines	250	∠	—	—	—
		Islamabad	Pakistan	10/250	∠	—	—	—
		Honiara	Solomon Is	5		—	—	—
		Mt Carlo	Monaco	100	—		—	
		Mahe	Seychelles	25	—	—		
		Kavalla	Greece	250	∠	—	—	
		Beyrouth	Lebanon	100				
9.550	31.41	Dar es Salaam	Tanzania	100	—	—	—	—
		Habana	Cuba	10/50	∠	—	—	—
		Makassar	Indonesia	7.5	∠	—	—	—
		Pori	Finland	15/250	—			—
		Fredrikstad	Norway	100	∠	—	—	
		Moskva	USSR	240/500	∠	—	—	
		Tachkent	USSR	100/500	∠			—
		Peking	China Rep					
		S M Galeria	Vatican	100				—
		Dacca	Bangladesh	7.5	∠			—
		Monrovia	Liberia	50	∠	—	—	—
		Male	Maldives	15		—	—	—
		Noblejas	Spain	350	/	—	—	—
		Bucuresti	Roumania	18/250	∠	—	—	—
		Allouis	France	100/500	∠	—	—	—
		Tokyo Kawagu	Japan	1	∠	—	—	—
		Shepparton	Australia	100	—	—	—	
		Bonaire Zuid	Neth Antilles	50				
		Santiago	Chile	100	/	—	—	
		Bombay	India	10	/			
9.555	31.40	London	UK	100/250	∠	—	—	—
		Tula	USSR	100	—			
		Irkutsk	USSR	100	∠	—	—	—
		Kenga	USSR	240	/	—		
		Mexico	Mexico	0.5/1	∠	—	—	—
		Tinang	Philippines	250	∠	—	—	—
		La Paz	Bolivia	10	∠	—	—	—

MHz	Metres	Station	Country	kW	M	M	S	N
		Okeechobee	USA	100				−
		Redwood City	USA	250	L	−	−	
		Salman Pack	Iraq	50	−	−	−	−
		Kimjae	Korea (S)	250	L	−	−	−
		Bata	Guinea	50	L	−	−	−
		Lisbonne	Portugal	50				−
		Biblis	Germany (W)	100				−
		Lampertheim	Germany (W)	20/100	L	−	−	−
		Kavalla	Greece	250	L	−	−	
9.560	31.38	Amman	Jordan	100	L	−	−	
		Serpukhov	USSR	240	L	−	−	−
		Krasnoiarsk	USSR	100	−			
		Allouis	France	100				
		Sofia	Bulgaria	50/500	L	−	−	−
		Meyerton	Australia	100	L	−	−	
		Carnarvon	Australia	100/250	L	−		
		Shepparton	Australia	100	−	−	−	
		Schwarzenburg	Switzerland	150	L	−	−	−
		Quito	Ecuador	100	−	−	−	−
		Tinang	Philippines	250	−	−	−	
		Noblejas	Spain	350	/			−
9.565	31.36	London	UK	250	L	−	−	
		Greenville	USA	50/250	L	−	−	
		Delano	USA	200	L	−	−	
		Tripoli	Libya	100	L	−	−	
		Kigali	Rwanda	250	L	−	−	−
		Simferopol	Ukraine	100				−
		Irkutsk	USSR	240		−	−	
		Frunze	USSR	50/500	L	−		−
		Recife	Brasil	10/15	L	−		−
		Julich	Germany (W)	100	L	−	−	
		Cyclops	Malta	250	L	−	−	−
		Holzkirchen	Germany (W)	10	L	−	−	
		Lisbonne	Portugal	50/250	L	−	−	−
		Sines	Portugal	250	−		−	
		Kimjae	Korea (S)	250		−		
		Aligarh	India	250	/			−
		Delhi	India	20	/			
		Pori	Finland	100/250			−	−
9.570	31.35	London	UK	250	L		−	−
		Kranji	Singapore	125	/	−	−	−
		Bucuresti	Roumania	18/250	L	−	−	−
		Doha	Qatar	100	L	−	−	−
		Shepparton	Australia	50	L	−	−	−
		Arganda	Spain	100	/	−	−	
		Noblejas	Spain	350	L	−		
		Jaji	Niger	10	L	−		
		Warszawa	Poland	30/100	L	−	−	−
		Bonaire Zuid	Neth Antilles	50	L	−	−	−

MHz	Metres	Station	Country	kW	M	M	S	N
		Malolos	Philippines	100	_			
		Santiago	Chile	10	L	_	_	_
			USSR					
9.575	31.33	Bombay	India	100	L	_	_	_
		Delhi	India	20/100	_	_	_	_
		Taipei	China Nat	10				
		Roma	Italy	60/100	L	_	_	_
		Godthaab	Greenland	10	L	_	_	_
		Armavir	USSR	240	L			_
		Tchita	USSR	100				
		Irkutsk	USSR	100	L	_	_	
		Vinnitsa	Ukraine	240	_			
		Pt Moresby	Papua New Guinea	10	L	_	_	_
		Mt Carlo	Monaco	100	/		_	_
		Kimjae	Korea (S)	100			_	
		Kamalabad	Iran	350	_			
		Pori	Finland	100/250	/		_	_
		Sackville	Canada	250				_
			Philippines					
		Sulaibiyah	Kuwait					
9.580	31.32	London	UK	100/250	L	_	_	_
		Ascension	Ascension	125/250	L	_	_	_
		Kranji	Singapore	250	/	_	_	_
		Limassol	Cyprus	100	_	_	_	
		Kazan	USSR	200	L	_	_	_
		Blagovechtchen	USSR	100	L			_
		Starobelsk	Ukraine	240/500	L			
		Kiev	Ukraine					
		Malolos	Philippines	50	_	_	_	_
		Shepparton	Australia	50/100	L	_	_	_
		Lusaka	Zambia	50	L	_	_	_
		Sackville	Canada	50	_	_	_	_
		Greenville	USA	50	L	_	_	_
		Arganda	Spain	100	/			
		Noblejas	Spain	350	_	_	_	_
		Suwon	Korea (S)	50	L	_	_	_
		Tanger	Morocco					
		Roma	Italy	60/100	/	_	_	_
9.585	31.30	S Paulo	Brasil	7.5/50	L	_	_	_
		Mogadiscio	Somalia	5	L	_	_	_
		Bizam Radio						
		Meyerton	South Africa	250/500	L	_	_	_
		Jaszbereny	Hungary	250		_	_	
		Tinang	Philippines	250				_
		Diosd	Hungary	100	L	_	_	_
		Tokyo Yamata	Japan	100	L		_	_
		Quito	Ecuador	100/500	_	_	_	_

MHz	Metres	Station	Country	kW	M	M	S	N
		V of Communist Party of Turkey						
		Bata	Guinea	50	∟	_	_	_
		Kavalla	Greece	250			_	
		Wien	Austria	100	/	_	_	_
		Sines	Portugal	250	_	_	_	
		Dacca	Bangladesh	100				
		Franceville	Gabon Rep	500			_	
		Pori	Finland	100				_
9.590	31.28	Montserrat	Br W Indies	15/50	_	_	_	_
		Masirah	Oman	100	/		_	_
		Kranji	Singapore	250			_	_
		Limassol	Cyprus	7.5/100	∟	_	_	_
		Bucuresti	Roumania	120/250	∟	_	_	
		Bonaire Noord	Neth Antilles	300	∟	_	_	_
		Aligarh	India	250	_	_	_	_
		Delhi	India	100	_	_	_	_
		Madras	India	10	/			
		Kazan	USSR	240	_	_		
		Irkutsk	USSR	240	_			
		Leningrad	USSR	100	∟			_
		Omsk	USSR	240	∟	_	_	_
		Starobelsk	Ukraine	240	/	_	_	
		Wertachtal	Germany (W)	500	∟	_	_	
		Julich	Germany (W)	100	/	_		
		Fredrikstad	Norway	100/250	/	_	_	
		S Domingo	Dominican Rep	0.3	∟	_	_	_
		Manzini	Swaziland	25/250		_	_	_
		Mt Carlo	Monaco	100	/		_	
		Kathmandu	Nepal	100	∟		_	
		Hoerby	Sweden	350	_		_	_
		Malolos	Philippines	50/100	/		_	_
		Sackville	Canada	50/250	/	_	_	_
		Islamabad	Pakistan	10				
		V of Front for Redemption of Somalia						
9.595	31.27	Montevideo	Uruguay	20	∟	_	_	_
		Biblis	Germany (W)	100	∟	_	_	_
		Lisbonne	Portugal	50/100	∟	_	_	_
		Tokyo Nagara	Japan	50	∟	_	_	_
		Salvador	Brasil	10	∟	_	_	_
		Bucuresti	Roumania	18/120	/	_	_	_
		Noblejas	Spain	350	_	_	_	_
		Dar es Salaam	Tanzania	50	_	_	_	_
		Kimjae	Korea (S)	100			_	
		R Kulmis	USSR					
9.600	31.25	London	UK	100/250	∟	_	_	_
		Ascension	Ascension	250	∟	_	_	_

101

MHz	Metres	Station	Country	kW	M	M	S	N
		Limassol	Cyprus	100			_	_
		Praha	Czechoslovakia	250	∟	_	_	_
		Mexico	Mexico	1	∟	_	_	_
		K Wusterhausen	Germany (E)	100	∟			_
		Sorong	Indonesia	5	_			
		Moskva	USSR	240	∟	_		
		Tachkent	USSR	50	∟	_	_	_
		Kenga	USSR	240		_	_	
		Okhotsk	USSR	50	∟	_	_	_
		Redwood City	USA	250	/			
		Greenville	USA	500				_
		Shepparton	Australia	100	∟	_	_	_
		Noblejas	Spain	350	∟	_	_	_
		Taipei	China Nat					
9.605	31.23	Limassol	Cyprus	100	_		_	_
		Allouis	France	100/500	_	_	_	_
		Stapok	Malaysia	10			_	
		Malolos	Philippines	100	∟	_	_	_
		Tinang	Philippines	50	_	_	_	_
		Brasilia	Brasil	10	/		_	
		Serpukhov	USSR	50	∟	_	_	_
		Duchanbe	USSR	500				
		S Cruz	Bolivia	5	∟	_	_	_
		Praha	Czechoslovakia	120	∟	_	_	_
		Wertachtal	Germany (W)	500	∟	_	_	_
		Julich	Germany (W)	100	∟	_	_	_
		Sackville	Canada	250	∟	_	_	_
		Tokyo Yamata	Japan	100	∟	_	_	_
		Mahe	Seychelles	30/100	_			
		S M Galeria	Vatican	100	∟	_	_	_
		Hoerby	Sweden	350	∟	_	_	_
		Karlsborg	Sweden	350			_	_
		Dacca	Bangladesh	100	_	_		
		Quito	Ecuador	100				
		Fredrikstad	Norway	120/250	_	_	_	_
		Riyadh	Saudi Arabia	350	∟	_	_	_
		Cyclops	Malta	250	/	_	_	
		Sines	Portugal	250	∟	_	_	
		Wien	Austria	100	_	_		
9.610	31.22	Limassol	Cyprus	100	/			
		Gedja	Ethiopia	10/100	_			
		Fredrikstad	Norway	100	_	_		_
		Julich	Germany (W)	100	∟	_	_	_
		Wertachtal	Germany (W)	500	∟	_	_	_
		Alma Ata	USSR	240/500	/		_	_
		Armavir	USSR	240/500				_
		Tula	USSR	240	∟	_	_	
		Khabarovsk	USSR	240	∟	_	_	_
		Nouakchott	Mauretania	30				_

MHz	Metres	Station	Country	kW	M	M	S	N
		Perth	Australia	10/50	∟	_	_	_
		Rio de Janeiro	Brasil	10	∟	_	_	_
		Brazzaville	Congo	50	∟	_	_	_
		Alger	Algeria	40/100	∟	_	_	_
		Sines	Portugal	250				
		Cyclops	Malta	250	∟	_	_	_
		Wavre	Belgium	250				
		Allouis	France	100	_	_		
		Mt Carlo	Monaco	100	∟	_	_	_
		Kimjae	Korea (S)	250				
		Athinai	Greece	100				_
		Meyerton	South Africa	500	/		_	_
		Bonaire Zuid	Neth Antilles	50				_
9.615	31.20	London	UK	100	/	_	_	_
		Limassol	Cyprus	20	/		_	_
		Aligarh	India	250	_	_	_	
		Delhi	India	100	∟	_	_	_
		Redwood City	USA	50	∟	_	_	_
		S Jose	Costa Rica	3/50	∟	_	_	_
		Marulas	Philippines	2.5	∟	_	_	_
		Tinang	Philippines	250		_	_	_
		Tanger	Morocco	50	/			_
		Rabat	Morocco	50/100	/			
		S M Galeria	Vatican	100	∟	_	_	_
		S Gabriel	Portugal					
		Sines	Portugal	250	∟	_	_	_
		Julich	Germany (W)	100				_
		Kavalla	Greece	250	∟	_	_	_
		Careysburg	Liberia	250				_
		Dacca	Bangladesh	7.5	_	_	_	
		Wavre	Belgium	100	/		_	_
		Mahe	Seychelles	25		_		
		Gedja	Ethiopia	100	/	_	_	_
		Allouis	France	100				_
		Alger	Algeria					
9.620	31.19	London	UK	250	/		_	_
		Limassol	Cyprus	100	/			
		Belgrade	Yugoslavia	100	∟	_	_	_
		Armavir	USSR	240	∟		_	
		Moskva	USSR	240	∟	_	_	_
		Vladivostock	USSR	100	∟	_	_	_
		Montevideo	Uruguay	20	∟	_	_	_
		Ho Chi Minh City	Vietnam	50				
		Maputo	Mozambique	10	∟	_	_	_
		Leipzig	Germany (E)	100			_	_
		Abu Dhabi	United Arab Emirates	120/ 250	∟	_	_	_
		Greenville	USA	50/250	∟	_	_	_
		Ismaning	Germany (W)	100	∟	_	_	

MHz	Metres	Station	Country	kW	M	M	S	N
		Quito	Ecuador	100/500	L	_	_	_
		V of Malayan Revln						
		Abu Zaabal	Egypt	100			_	
		Wien	Austria	100	_	_	_	_
		Bonaire Zuid	Neth Antilles	50	_	_		
		Wellington	New Zealand	7.5	_	_		
		S Gabriel	Portugal	100	/	_	_	_
		Allouis	France	100				_
9.625	31.17	London	UK	100				_
		Sackville	Canada	50/250	L	_	_	_
		Aligarh	India	100				_
		Delhi	India	20/50	/			
		S M Galeria	Vatican	100	L	_	_	_
		Schwarzenburg	Switzerland	100	_			
		Bucuresti	Roumania	18/120	L	_	_	_
		Moskva	USSR	240				_
		Tbilisi	USSR	240		_	_	
		Tchita	USSR	240	L			_
		Tinang	Philippines	250		_	_	_
		Peking	China Rep					
		Biblis	Germany (W)	20/	L	_		
				100				
		Lampertheim	Germany (W)	100			_	
		Playa de Pals	Spain	100	/			_
		Greenville	USA	50/250		_	_	_
		Redwood City	USA	250			_	
		Mt Carlo	Monaco	100	_			
		Ho Chi Minh City	Vietnam					
9.630	31.15	Noblejas	Spain	350/700	L	_	_	_
		Delhi	India	20/100	L	_	_	_
		Velkekostolany	Czechoslovakia	120	L	_	_	_
		Moskva	USSR	50		_		
		Serpukhov	USSR	200	L		_	_
		Roma	Italy	60/100				_
		Tinang	Philippines	250	L	_	_	_
		Karlsborg	Sweden	350	L	_	_	_
		Jerusalem	Israel	50/300	L	_	_	_
		Mt Carlo	Monaco	100	_	_		
		Santiago	Chile	10	L	_	_	_
		Tanger	Morocco	100	/	_	_	
		Bonaire Noord	Neth Antilles	300	L	_	_	
		Careysburg	Liberia	250				_
		Okeechobee	USA	100				
		Julich	Germany (W)	100	L	_	_	_
		Lopik	Netherlands	100	/			
		Riyadh	Saudi Arabia	350			_	_
9.635	31.14	London	UK	100/250	L	_	_	_
		Aparacida	Brasil	7.5	L	_	_	_
		Bamako	Malawi	18			_	_

MHz	Metres	Station	Country	kW	M	M	S	N
		Greenville	USA	250/500	∠	_	_	_
		Singapore	Singapore	50	∠	_	_	_
		Ivanofrankovsk	Ukraine	240				_
		Kaunas	USSR	50		_	_	
		Vladivostock	USSR	50	∠	_	_	_
		Bogota	Colombia	25	∠	_	_	_
		Babel	Iraq	500	_	_		
		Beira	Mozambique	100	∠	_	_	_
		Hoerby	Sweden	350	_			
		Sines	Portugal	250	∠	_	_	_
		S Gabriel	Portugal	100	_	_	_	
		Sackville	Canada	250				
		Quito	Ecuador					
9.640	31.12	London	UK	100/250	∠	_	_	_
		Antigua	Br W Indies	250	∠	_	_	_
		Montserrat	Br W Indies	50	_			
		Limassol	Cyprus	7.5/100	∠	_	_	_
		Greenville	USA	250/500	∠		_	
		Julich	Germany (W)	100	∠	_	_	_
		Wertachtal	Germany (W)	500	∠	_	_	_
		Montevideo	Uruguay	10	∠	_	_	_
		Moskva	USSR	240	∠	_	_	_
		Bucuresti	Roumania	120	/		_	
		Suwon	Korea (S)	50	∠	_	_	_
		Mt Carlo	Monaco	100	_	_		
		Caracas	Venezuela	10	∠	_	_	_
		Carnarvon	Australia	100	∠	_	_	_
		Alger	Algeria	50	∠	_	_	_
		Malolos	Philippines	100	_			
		Kavalla	Greece	250	/			
		Athinai	Greece	100	/			
9.645	31.10	London	UK	100/250		_		
		Fredrikstad	Norway	100	∠	_	_	_
		Karachi	Pakistan	10/100	∠	_	_	_
		Islamabad	Pakistan	100/250	_		_	_
		S Paulo	Brasil	7.5	∠	_	_	_
		Pocas Caldas	Brasil	7.5	∠	_	_	_
		S Jose	Costa Rica	1	∠	_	_	_
		S M Galeria	Vatican	100	∠	_	_	_
		Novosibirsk	USSR			_	_	
		Khabarovsk	USSR	50	∠	_		_
		Nauen	Germany (E)	500	_	_		
		Delhi	India	50	_	_	_	_
		Malolos	Philippines	100	∠	_	_	_
		Tinang	Philippines	250	_	_	_	
		Pori	Finland	100	_			
		Wavre	Belgium	100	/			
			Australia					
9.650	31.09	Greenville	USA	50/500	∠	_	_	_

105

MHz	Metres	Station	Country	kW	M	M	S	N
		Delano	USA	250	/		_	_
		Dixon	USA	250		_	_	
		Conakry	Guinea	100	∠			
		Baku	USSR	100			_	_
		Moskva	USSR	240	∠	_	_	_
		Montevideo	Uruguay	10	∠	_	_	_
		Poro	Philippines	50	/			_
		Tinang	Philippines	50/250	∠	_	_	_
		Ismaning	Germany (W)	100	∠	_	_	
		Julich	Germany (W)	100	∠	_	_	
		Sines	Portugal	250	∠	_	_	_
		Magwa	Kuwait	50/250	∠	_	_	_
		Cyclops	Malta	250	∠	_	_	_
		Wien	Austria	100	_	_	_	
		Tanger	Morocco	100	∠	_	_	
		Santiago	Chile	10	_			
		Mahe	Seychelles	25	_	_		
		Tripoli	Libya					
		Kimjae	Korea (S)	250			_	_
		Meyerton	South Africa	250		_		
		R of the Patriots						
		Mt Carlo	Monaco					
		Quito	Ecuador	100	/			
		Jerusalem	Israel					
9.655	31.07	Patumthani	Thailand	100	∠	_	_	_
		Tripoli	Libya	100	∠	_	_	_
		Sackville	Canada	50/250	∠	_	_	_
		Jaszbereny	Hungary	250	/	_	_	_
		Diosd	Hungary	100	∠			_
		Armavir	USSR	100	∠			_
		Frunze	USSR	240			_	_
		Orcha	Bielorussia	100			_	_
		Mt Carlo	Monaco	100	/			
		Habana	Cuba	10/50			_	_
		Athinai	Greece	100	∠	_	_	_
		Kavalla	Greece	250	_	_	_	_
		Thessaloniki	Greece	35	/			
		Bogota	Colombia	10	∠	_	_	_
		Wavre	Belgium	100/250	_	_	_	_
		Lhasa	Tibet					
9.660	31.06	London	UK	250	∠	_	_	_
		Luanda	Angola	100	/			
		Brisbane	Australia	10	∠	_	_	_
		Tachkent	USSR	240	_			
		Kavalla	Greece	250	∠	_	_	_
		Dixon	USA	250	∠	_	_	_
		Okeechobee	USA	100			_	
		Delano	USA	250				_
		Tanger	Morocco	100		_		

MHz	Metres	Station	Country	kW	M	M	S	N
		Kinshasa	Zaire	50	∟	_	_	_
		Islamabad	Pakistan	100	∟	_	_	
		Karachi	Pakistan	10/250		_	_	_
		Lopik	Netherlands	100		_	_	_
		Karlsborg	Sweden	350	/		_	_
		Hoerby	Sweden	350	/	_	_	_
		Allouis	France	100/500	∟	_	_	_
		Lampertheim	Germany (W)	100		_	_	
		Biblis	Germany (W)	100	∟			_
		Playa de Pals	Spain	250		_	_	
		Pori	Finland	100	_			
		Sorong	Indonesia	5	/	_	_	_
		Wien	Austria	100	_			
		Mt Carlo	Monaco	100				
		Sines	Portugal	250	_			
		Schwarzenburg	Switzerland	150/250	/			_
		Kimjae	Korea (S)	100				_
		Tripoli	Libya					
		Caracas	Venezuela					
9.665	31.04	Kajang	Malaysia	50	∟	_	_	_
		Brasilia	Brasil	7.5/10	/			_
		Nairobi	Kenya	100	∟	_	_	_
		Nauen	Germany (E)	500	/			_
		K Wusterhausen	Germany (E)	50	∟			_
		Leipzig	Germany (E)	100	_			_
		Voronej	USSR	240	_			_
		Kinghisepp	USSR	240		_		
		Tchita	USSR	100	∟			_
		Ivanofrankovsk	Ukraine	240	∟	_	_	_
		Vinnitsa	Ukraine	100		_		
		Karlsborg	Sweden	350		_	_	
		Hoerby	Sweden	350		_	_	
		Suwon	Korea (S)	50		_	_	
		Ankara	Turkey	250	/		_	_
		Allouis	France	100		_		_
		Quito	Ecuador	100	/		_	_
		Sines	Portugal	250	/		_	_
9.670	31.02	Greenville	USA	50/500	∟	_	_	_
		Montevideo	Uruguay	10	∟	_	_	_
		Jeddah	Saudi Arabia	50/100	∟	_	_	_
		Irkutsk	USSR	100	∟	_	_	_
		Kiev	Ukraine	100	∟	_	_	_
		Sines	Portugal	250	∟	_	_	_
		Peking	China Rep					
		Careysburg	Liberia	250	/	_	_	_
		Tinang	Philippines	250			_	
		Cyclops	Malta	250	/	_	_	

MHz	Metres	Station	Country	kW	M	M	S	N
		Tokyo Yamata	Japan	200	∠	_	_	_
		Shepparton	Australia	10/100	∠	_	_	_
9.675	31.01	Florianapolis	Brasil	10	∠	_	_	_
		Tokyo Yamata	Japan	50/100	∠	_	_	_
		Armavir	USSR	100/240	∠	_	_	_
		Voronej	USSR	240	∠	_		_
		Novosibirsk	USSR	100			_	
		Simferopol	Ukraine	240	_			_
		Warszawa	Poland	100	∠	_	_	_
		Aligarh	India	250	∠	_	_	_
		Delhi	India	20/100	∠	_	_	_
		Abu Zaabal	Egypt	100				
		Mt Carlo	Monaco	100	∠	_	_	_
		Kimjae	Korea (S)	50	_	_		
		Athinai	Greece	100	∠	_	_	
		Santiago	Chile	45	_			
		Arganda	Spain	100	/		_	_
		Sines	Portugal	250	/		_	_
9.680	30.99	London	UK	250	∠	_	_	_
		Lyndhurst	Australia	10	∠	_	_	_
		Meyerton	South Africa	100	∠	_	_	_
		Montevideo	Uruguay	10	∠	_	_	_
		Tchita	USSR	100		_	_	
		Kavalla	Greece	250	∠	_	_	
		Jakarta	Indonesia	50	∠	_	_	
		Quito	Ecuador	100	∠	_	_	_
		Delano	USA	250	_			
		Greenville	USA	250	∠	_	_	_
		Ismaning	Germany (W)					
		Lampertheim	Germany (W)	100		_	_	
		Holzkirchen	Germany (W)	10				_
		Biblis	Germany (W)	100	_			
		Playa de Pals	Spain	100	∠	_	_	_
		Wavre	Belgium	100	_			
		Alger	Algeria	100	∠	_	_	_
		Dacca	Bangladesh	100			_	
		Tanger	Morocco	100				_
9.685	30.98	Alger	Algeria	50/100	∠	_	_	_
		Erevan	USSR	100	/			
		Moskva	USSR	240/500	∠	_	_	_
		Irkutsk	USSR	240/500	∠		_	
		Lvov	Ukraine	100		_	_	
		Panama	Panama	1	∠	_	_	_
		S Paulo	Brasil	7.5	/			_
		Habana	Cuba	50	∠		_	
		Bucuresti	Roumania	120	∠			_
		Kampala	Uganda	250			_	
		Schwarzenburg	Switzerland	150	_			
		Wavre	Belgium	100	/	_	_	_

MHz	Metres	Station	Country	kW	M	M	S	N
		Arganda	Spain	100	∠	_	_	_
		Taipei	China					
		Redwood City	USA	250		_		
		Cincinnati	Portugal	175	∠	_	_	_
		Sines	Portugal			_		
		Santiago	Chile	100	/		_	
		Quito	Ecuador					
		Sofia	Bulgaria	500				
9.690	30.96	London	UK	250	∠	_	_	
		Antigua	Br W Indies	250	∠	_	_	
		Limassol	Cyprus	20/100	∠	_	_	_
		Julich	Germany (W)	100	∠	_	_	_
		Tula	USSR	100	∠	_	_	_
		Gral Pacheo	Argentina	100	∠	_	_	_
		Islamabad	Pakistan	100/250	_			
		Karachi	Pakistan	50	∠	_	_	
		Okeechobee	USA	100	∠		_	_
		Redwood City	USA	250	/			
		Scituate	USA	50	_			
		Bucuresti	Roumania	18/250	∠	_	_	
		Santiago	Chile	10	_			
		Malolos	Philippines	50/100	_			
		Noblejas	Spain	350	_			
		Kavalla	Greece	250	∠	_	_	_
		Allouis	France	100/500	/	_	_	_
		Hoerby	Sweden	350	_			
		Karlsborg	Sweden	350			_	_
		Pt au Prince	Haiti	100				_
9.695	30.94	Limassol	Cyprus	100		_		
		Manaos	Brasil	7.5	∠	_	_	_
		Holzkirchen	Germany (W)	10		_		
		Biblis	Germany (W)	100	∠	_	_	_
		Lisbonne	Portugal	50/250	∠	_	_	_
		Kazan	USSR	240	_		_	
		Frunze	USSR	100			_	
		Petropavlo Kam	USSR	100	/			
		Tbilisi	USSR	240	_			
		Karlsborg	Sweden	350	/		_	_
		Hoerby	Sweden	350	∠	_	_	_
		Allouis	France	100/500	_	_	_	_
		Noblejas	Spain	350	∠	_	_	
		Rarotonga	Cook Is	0.5		_	_	
		Franceville	Gabon Rep	500			_	
		Phnom Penh	Cambodia					
9.700	30.93	London	UK	250			_	
		Ascension	Ascension	125/250	_		_	
		Dixon	USA	100	∠	_	_	_
		Sofia	Bulgaria	50/145	∠	_	_	_
		Tchita	USSR	50	/	_	_	_

MHz	Metres	Station	Country	kW	M	M	S	N
		Tula	USSR	50	_			
		Kazan	USSR	100		_		
		Kenga	USSR	50	/			
		Carnarvon	Australia	250	_	_		_
		Julich	Germany (W)	100	L	_	_	_
		Kigali	Rwanda	250	L	_	_	_
		Tokyo Yamata	Japan	100/200	L	_	_	_
		Kavalla	Greece	250	L	_	_	_
		Malolos	Philippines	100	_			
		Careysburg	Liberia	250		_		
		Schwarzenburg	Switzerland	250	_			_
9.705	30.91	London	UK	100	/			
		Limassol	Cyprus	100			_	
		Gedja	Ethiopia	100		_	_	
		Aligarh	India	250	L	_	_	_
		Delhi	India	20/100	/			_
		Holzkirchen	Germany (W)	10			_	
		Biblis	Germany (W)	100	L	_		
		Lampertheim	Germany (W)	100				
		Playa de Pals	Spain	250			_	
		Lisbonne	Portugal	50	L	_	_	_
		Tokyo Yamata	Japan	20	L	_	_	_
		Rio de Janeiro	Brasil	7.5	L	_	_	_
		Mexico	Mexico	10	L	_	_	_
		Lvov	Ukraine	240	_			_
		Alger	Algeria	50	L	_	_	_
		Sofia	Bulgaria					
		Allouis	France	100/500	_	_		_
		Abis	Egypt	250			_	
		Quito	Ecuador					
9.710	30.90	S Fernando	Argentina	6	L	_	_	_
		Malherbes	Mauritius	10				_
		Mauritius	Mauritius	10	/		_	_
		Kaunas	USSR	100	/	_	_	_
		Kiev	Ukraine	240	L	_	_	
		Komsomolskamur	USSR	240	_			
		Frunze	USSR	240/500	L	_		_
		Vladivostock	USSR	50				_
		Petropavlo Kam	USSR	100		_	_	
		Roma	Italy	60/100	L	_	_	_
		Penang	Malaysia	10	L	_	_	
		Meyerton	South Africa	100	L	_	_	
		Thessalonika	Greece	35	_	_	_	
		Koebenhaven	Denmark	50		_		
		Kimjae	Korea (S)	250	/		_	_
9.715	30.88	London	UK	100/250	_	_	_	_
		Limassol	Cyprus	20				
		Quito	Ecuador	100	L	_		
		Bocaue	Philippines	10/50	L	_	_	_

110

MHz	Metres	Station	Country	kW	M	M	S	N
		Iba	Philippines	100	/			
		Lopik	Netherlands	100	∠	_	_	_
		Brazzaville	Congo	50	∠	_	_	_
		Kiev	Ukraine	100/240				_
		Orcha	Bielorussia	100	∠	_	_	_
		Sucre	Bolivia	2	∠	_	_	_
		Bonaire Noord	Neth Antilles	300	∠	_	_	_
		Tanger	Morocco	50	∠	_	_	_
		Greenville	USA	250				_
		Okeechobee	USA	100	_			
		Scituate	USA	50/100	∠	_	_	_
		Allouis	France	100/500	∠	_	_	_
		Sackville	Canada	50/250	_			
		Ismaning	Germany (W)	100	∠	_		
		Lampertheim	Germany (W)	100				
		Playa de Pals	Spain	100				_
		Careysburg	Liberia	250	/		_	_
		Santiago	Chile	100	/		_	_
		Madras	India	100	/			
		Sines	Portugal	250	/			
9.720	30.86	Ekala	Ceylon	10/35	∠	_	_	_
		Riazan	USSR	240/500	∠	_	_	_
		Kiev	Ukraine	240	_	_		
		Vinnitsa	Ukraine	240/500	/		_	_
		Diriyya	Saudi Arabia	50	∠	_	_	_
		Kimjae	Korea (S)	100/250	∠	_	_	_
		Kamalabad	Iran	100	_			_
		Dacca	Bangladesh					
		Hyderabad	India	10	/			
9.725	30.85	Kranji	Singapore	250	∠	_	_	_
		Nikolaevskamur	USSR	50	_	_		_
		Greenville	USA	50	_			
		Schwarzenburg	Switzerland	150	∠	_	_	_
		Tinang	Philippines	250	∠	_	_	_
		Noblejas	Spain	350	∠	_	_	_
		Mt Carlo	Monaco	100	_	_		
		Wien	Austria	100	∠	_	_	_
		Lisbonne	Portugal	100	∠	_	_	_
		Holzkirchen	Germany (W)	10	/	_		
		Biblis	Germany (W)	100	/		_	_
		Lampertheim	Germany (W)	50				_
9.730	30.83	Dixon	USA	100	∠	_	_	_
		Nauen	Germany (E)	50	∠	_	_	_
		Leipzig	Germany (E)	50	∠	_	_	_
		Serpukhov	USSR	100	_	_		
		Tula	USSR	100	∠			_
		Riyadh	Saudi Arabia	350	_		_	
		Rangoon	Burma	50	∠	_	_	_
		Wavre	Belgium	100	_		_	_

MHz	Metres	Station	Country	kW	M	M	S	N
		Kampala	Uganda	250				
		S M Galeria	Vatican	100				—
		Delhi	India	100	_	_	_	_
		Mt Carlo	Monaco	100	_			—
		Sackville	Canada	50				—
		Manzini	Swaziland	25		_	_	_
		Quito	Ecuador	100				—
9.735	30.82	London	UK	100/250	L	_	_	_
		Antigua	Br W Indies	125/250	L	_	_	_
		Asuncion	Paraguay	3/100	L	_	_	_
		Wertachtal	Germany (W)	500	L	_	_	_
		Kigali	Rwanda	250	L	_	_	_
		Lvov	Ukraine	500				—
		Tula	USSR	240	_	_	_	
		Duchanbe	USSR	500				—
		Nikolaevskamur	USSR	240	L	_	_	_
		Novosibirsk	USSR	240/500	L			—
		Irkutsk	USSR	100/500	_	_		
		Mt Carlo	Monaco	100	/	_	_	_
		Cyclops	Malta	250	L	_	_	_
		Tanger	Morocco	100				—
		Kavalla	Greece	250				—
		Quetta	Pakistan	10	/			
9.740	30.80	Kranji	Singapore	250	/			
		Lomas Mirador	Argentina	10	L	_	_	_
		S Gabriel	Portugal	100	L	_	_	_
		Moskva	USSR	120/240			_	_
		Simferopol	Ukraine	100			_	_
		Careysburg	Liberia	250	L	_	_	
		Abis	Egypt	250				
		Velkekostolany	Czechoslovakia	120/200	L	_	_	_ _
		Malolos	Philippines	100	L	_	_	_
		Jerusalem	Israel	20/300	_		_	_
9.745	30.79	London	UK	100	_			
		Moskva	USSR	240	/	_	_	_
		Kenga	USSR	500	/			_
		Quito	Ecuador	30/100	L	_	_	_
		S Paulo	Brasil	1/7.5	L	_	_	_
		Jayapura	Indonesia	5	L	_	_	_
		Babel	Iraq	100/500	_	_		
		Salman Pack	Iraq	500			—	
		Hoerby	Sweden	350	_			—
		Greenville	USA	250	_	_		
		Sofia	Bulgaria	500			_	_
		Yaounde	Cameroon					
		Noblejas	Spain	350	/			—
9.750	30.77	London	UK	100/250	L	_	_	_
		Limassol	Cyprus	20/100	L	_	_	_

MHz	Metres	Station	Country	kW	M	M	S	N
		Careysburg	Liberia	250	∠	_	_	_
		Madras	India	100	∠	_	_	_
		Krasnoiarsk	USSR	100	_	_	_	
		Sverdlovsk	USSR	100	/	_	_	
		Petropavlo Kam	USSR	100		_	_	
		Tachkent	USSR	240			_	
		Vladivostock	USSR	100				_
		Komsomolskamur	USSR	50	∠			
		Simferopol	Ukraine	240		_	_	
		Papeete	Tahiti	4	∠	_	_	_
		Dar es Salaam	Tanzania	50	_			_
		Santiago	Chile	10	∠		_	_
		Tirane	Albania					
		Kajang	Malaysia		/			
		Dacca	Bangladesh	100	_	_		
		Bucuresti	Roumania	18/120	∠	_	_	_
		Playa de Pals	Spain	100	/			_
		Biblis	Germany (W)	100	∠	_		
		Holzkirchen	Germany (W)	10		_		
		Lampertheim	Germany (W)	100	∠		_	_
		Lisbonne	Portugal	250	_			
		Allouis	France	100	_		_	_
		Redwood City	USA	250	_			
9.755	30.75	Mokattam	Egypt	100		_		
		Goiana	Brasil	7.5	∠	_	_	_
		Duchanbe	USSR	100		_	_	
		Irkutsk	USSR	240		_		
		Nikolaevskamur	USSR	240	∠	_	_	
		K Wusterhausen	Germany (E)	100	_	_		
		Cincinnati	USA	175/250	_	_		
		Delhi	India	100	∠	_	_	_
		Warszawa	Poland	30/60	∠	_	_	_
		Wavre	Belgium	250	∠	_	_	_
		Bonaire Zuid	Neth Antilles	50	_			
		Iba	Philippines	100	∠	_	_	
		Bocaue	Philippines	50	_	_		
		Santiago	Chile	100	_			
		Pori	Finland	100		_	_	
		Sackville	Canada	50/250		_	_	
		Allouis	France	100	/			
		V of Kampuchean People						
9.760	30.74	London	UK	100/250	∠	_	_	_
		Ascension	Ascension	250	_			
		Tanger	Morocco	35/100	∠	_	_	_
		Ivanofrankovsk	Ukraine	240	_			
		Khabarovsk	USSR	240	/	_	_	_
		Kinghisepp	USSR	240	∠	_		
		Tbilisi	USSR	240	∠	_		_

MHz	Metres	Station	Country	kW	M	M	S	N
		Tokyo Nagara	Japan	10	L	_	_	_
		Hurlingham	Argentina	20	L	_	_	_
		Tirane	Albania					
		Tinang	Philippines	250	L	_	_	_
		Quito	Ecuador	100				
		Shepparton	Australia	100	L	_	_	_
		Athinai	Greece	100	L	_	_	_
		Kavalla	Greece	250	L	_	_	_
		Alger	Algeria	100	L	_	_	_
		Cairo	Egypt					
9.765	30.72	Ascension	Ascension	250		_	_	_
		Antigua	Br W Indies	125/250	/			_
		Julich	Germany (W)	100	L	_	_	_
		Wertachtal	Germany (W)	500	L	_	_	_
		Armavir	USSR	100		_	_	
		Kazan	USSR	100			_	
		Leningrad	USSR	240	/	_		_
		Lvov	Ukraine	240/500	L	_	_	_
		Greenville	USA	500	L	_	_	_
		Quito	Ecuador	100	L	_	_	_
		Kamalabad	Iran	100		_	_	
		Habana	Cuba	10				_
		Kimjae	Korea (S)	100	L	_	_	_
		Bocaue	Philippines	50	_			
		Schwarzenburg	Switzerland	500	/			
		Sottens	Switzerland	500	_			_
		Sofia	Bulgaria	500	_	_		
		Taipei	China Nat					
		Wien	Austria	100	_			
		Santiago	Chile	100	/			_
9.770	30.71	London	UK	100	L	_	_	_
		Limassol	Cyprus	100	/	_	_	
		Ascension	Ascension	250	_			
		Jakarta	Indonesia	20	L	_	_	
		Montevideo	Uruguay	10	L	_	_	_
		Rio de Janeiro	Brasil	7.5/10	L	_	_	_
		Serpukhov	USSR	100	/			
		Vladivostock	USSR	100	_			_
		Kenga	USSR	100	/			
		Wien	Austria	100	L	_	_	_
		Poro	Philippines	100	L	_	_	_
		Kinshasa	Zaire	10	L	_	_	_
		Bonaire Noord	Neth Antilles	300	L	_	_	_
		Kavalla	Greece	250	L	_	_	_
		Habana	Cuba	10	L	_	_	_
		Nauen	Germany (E)	100/500	/	_	_	_
		Tanger	Morocco	100	_		_	_

MHz	Metres	Station	Country	kW	M	M	S	N
		Wellington	New Zealand	7.5	_	_	_	
		Abu Zaabal	Egypt	100			_	
		Kamalabad	Iran	100	_			_
		Sanaa	Yemen	50			_	
		Lopik	Netherlands	100	_	_		
		Lampertheim	Germany (W) 50/100					_
		Shepparton	Australia	50/100	/	_	_	
		Cyclops	Malta	250	_	_		
		Rawalpindi	Pakistan	100	/			
9.775	30.69	Moskva	USSR	200				
9.780	30.67	Moskva	USSR					
		Tirane	Albania	240				
		Sanaa	Yemen					
9.785	30.66	Moskva	USSR	50/240				
9.790	30.64	Tirane	Albania	240				
		Moskva	USSR					
		Vladivostock	USSR					
		Karachi	Pakistan					
		Madrid	Spain					
9.795	30.63	Kazan	USSR	100				
		Minsk	USSR					
9.800	30.61	Moskva	USSR	100				
9.805	30.60	Cairo	Egypt	250				
9.810	30.58	Moskva	USSR	100				
9.815	30.57	Jerusalem	Israel					
			USSR					
9.820	30.55	Peking	China Rep					
		Jerusalem	Israel					
			USSR					
9.825	30.53	London	UK	100/250				
9.830	30.52		USSR					
9.833	30.51	Budapest	Hungary					
		Jerusalem	Israel					
9.840	30.49	Hanoi	Vietnam	100				
		Kuwait	Kuwait					
9.850	30.46	Cairo	Egypt	100				
		Moskva	USSR					
		Peking	China Rep					
9.858	30.43	London*	UK					
9.860	30.43	Peking	China Rep					
		Alger	Algeria					
9.870	30.40	Seoul	Korea (S)					
9.875	30.38	Julich*	Germany (W)					
9.880	30.36	Peking	China Rep					
9.887	30.34	Hanoi	Vietnam					
		Riyadh	Saudi Arabia					
9.893	30.32	Peking	China Rep					
9.895	30.32	Hilversum	Netherlands					
			China Rep					

115

MHz	Metres	Station	Country	kW
9.900	30.30	Peking	China Rep	
9.912	30.27	Delhi	India	100
9.915	30.26	London	UK	
9.920	30.24	Peking	China Rep	
9.929	30.21	(AFRTS)*	USA	
9.940	30.18	Peking	China Rep	
9.945	30.17	Peking	China Rep	
9.950	30.15	Delhi	India	
9.953	30.14	R Freedom for South Yemen		
9.965	30.11	Peking	China Rep	
9.977	30.07	Pyongyang	Korea (N)	
9.985	30.05	V of NUFC		
9.995	30.02	Cairo	Egypt	
10.000	30 00	Rugby Freq Std	UK	
		Honolulu Freq Std	Hawaii	
		Boulder Freq Std	USA	
10.009	29.98		USSR	
10.010	29.97	Kathmandu	Nepal	
		Hanoi	Vietnam	
10.040	29.88	Hanoi	Vietnam	
10.055	29.84	V of Iraqi Kurdistan		
10.060	29.82	Hanoi	Vietnam	
10.080	29.76	V of NUFK		
10.120	29.64	V of NUFC		
		Moskva*	USSR	
10.125	29.63			
10.190	29.44	R Liberty*		
10.225	29.34	Hanoi	Vietnam	
10.235	29.31	Greenville*	USA	
10.245	29.28	Peking	China Rep	
10.260	29.24	Peking	China Rep	
10.315	29.08	RFE/R Liberty*		
10.335	29.03	Delhi	India	
10.338	29.02	Moskva*	USSR	
10.356	28.97	Allouis	France	
10.380	28.90	Greenville*	USA	
10.385	28.89		Saudi Arabia*	
10.420	28.79	RFE/R Liberty*		
10.454	28.70	Greenville*	USA	
10.537	28.47		USA*	
10.615	28.26			
10.620	28.25		USSR*	
10.660	28.14		USSR*	
10.690	28.06		USSR*	
10.695	28.05		USSR*	
10.740	27.93	Moskva*	USSR	
10.761	27.88		Germany (W)*	
10.855	27.64		USSR*	

MHz	Metres	Station	Country	kW
10.865	27.61	Peking	China Rep	
10.869	27.60	Bethany*	USA	
		Greenville*	USA	
10.880	27.57	Monrovia*	Liberia	
10.905	27.51	Lisbon*	Portugal	
10.922	27.47		Germany (W)*	
10.972	27.34	Tangier*	Morocco	
11.000	27.27	Peking	China Rep	
11.033	27.19	Paris*	France	
11.040	27.17	Peking	China Rep	
11.100	27.03	Peking	China Rep	
11.290	26.57	Peking	China Rep	
11.300	26.55	Peking	China Rep	
11.330	26.48	Peking	China Rep	
11.340	26.46	Moskva	USSR	
11.350	26.43	Pyongyang	Korea (N)	
11.375	26.37	Peking	China Rep	
11.445	26.21	Peking	China Rep	
11.455	26.19	Peking	China Rep	
11.495	26.10		USSR*	
11.500	26.09	Peking	China Rep	
		Riyadh	Saudi Arabia	
11.505	26.08	Peking	China Rep	
11.515	26.05	Peking	China Rep	240
11.530	26.02	Peking	China Rep	
11.533	26.01	Pyongyang	Korea (N)	
11.568	25.93	Pyongyang	Korea (N)	
11.575	25.92	RFE/R Liberty		
			USSR*	
11.585	25.90		USSR	
11.600	25.86	Peking	China Rep	120
		V of Democratic Kampuchea		
11.602	25.85	Jerusalem	Israel	
11.610	25.84	Peking	China Rep	
11.620	25.82	Delhi	India	
		Jerusalem	Israel	
11.625	25.81	Jerusalem	Israel	
11.630	25.80	Cairo	Egypt	100
		Moskva	USSR	240
		Peking	China Rep	
11.635	25.78	Peking	China Rep	
		Jerusalem	Israel	
11.640	25.77	Karachi	Pakistan	
		Jerusalem	Israel	
11.645	25.76	Hargeisa	Somalia	5
11.650	25.75	Peking	China Rep	
		Dacca	Bangladesh	
11.655	25.74	Jerusalem	Israel	300

117

MHz	Metres	Station	Country	kW	M	M	S	N
11.660	25.73	Peking	China Rep					
		V of Democratic Kampuchea						
11.665	25.72	Peking	China Rep					
11.670	25.71		USSR					
11.672	25.70	Karachi	Pakistan					
11.675	25.70	Peking	China Rep	240				
		RFE*						
11.680	25.68	London	UK					
		Karachi	Pakistan					
11.685	25.67	Peking	China Rep					
		Riyadh	Saudi Arabia					
11.690	25.66	Moskva	USSR					
		Peking	China Rep					
		Kiev	Ukraine					
11.695	25.65	Peking*	China Rep	120/240				
11.700	25.64		USSR	50				
		Vatican	Vatican					
		Berlin	Germany	250				
		S Domingo	Dominican Rep	50				
		Tripoli	Libya					
		R of the Patriots						
		Jerusalem	Israel					
11.705	25.63	Limassol	Cyprus	100	/			
		Montserrat	Br W Indies	50	/			_
		Okeechobee	USA	100			_	
		Simferopol	Ukraine	500	/			
		Khabarovsk	USSR	100		_	_	
		Serpukhov	USSR	240	_	_	_	_
		Karlsborg	Sweden	350	_	_	_	_
		Hoerby	Sweden	350	/	_	_	_
		Nauen	Germany (E)	50	_	_	_	_
		K Wusterhausen	Germany (E)	100	_		_	
		Tokyo Yamata	Japan	100	_	_	_	_
		S M Galeria	Vatican	100	_	_	_	_
		Wellington	New Zealand	7.5	_			
		Mt Carlo	Monaco	100	_	_	_	_
		Allouis	France	100/500	_	_	_	_
		Jerusalem	Israel	300	_		_	_
		Shepparton	Australia	50/100	_	_	_	_
		Carnarvon	Australia	100	_	_	_	
		Sackville	Canada	250	_			
		Santiago	Chile	25/100	_			
		Islamabad	Pakistan	100		_	_	_
		Karachi	Pakistan	50	_			
		Mahe	Seychelles	100	_	_	_	
		Careysburg	Liberia	250				
		Bucuresti	Roumania	250		_	_	
		Bonaire Zuid	Neth Antilles	50		_	_	

MHz	Metres	Station	Country	kW	M	M	S	N
		Peking	China Rep					
		Julich	Germany (W)	100				
11.710	25.62	London	UK	250	∠	_	_	_
		Limassol	Cyprus	100		_	_	_
		Kranji	Singapore	250	∠		_	_
		Gral Pacheco	Argentina	100	∠		_	_
		Armavir	USSR	240	∠			_
		Frunze	USSR	50		_	_	
		Kazan	USSR	100	/			
		Komsomolskamur	USSR	100	_			
		Tbilisi	USSR	240			_	_
		Noumea	New Caledonia	4	∠	_	_	_
		Tanger	Morocco	35/100				_
		Bucuresti	Roumania	120	∠	_		_
		Bonaire Zuid	Neth Antilles	50	/	_	_	
		Manzini	Swaziland	25			_	
		Careysburg	Liberia	250				_
		Peking	China Rep					
		Mt Carlo	Monaco					
11.715	25.61	London	UK	100	∠	_	_	
		Schwarzenburg	Switzerland	150	∠	_	_	_
		Delhi	India	20/100	_	_	_	_
		Jakarta	Indonesia	100/120	∠	_	_	_
		Orcha	Bielorussia	240		_	_	
		Komsomolskamur	USSR	50	/			
		Duchanbe	USSR	240/500	∠	_	_	_
		Sverdlovsk	USSR	100	/			
		Poro	Philippines	50	_		_	_
		Tinang	Philippines	250	/	_		_
		Abis	Egypt	250		_		
		S M Galeria	Vatican	100	∠	_	_	_
		Greenville	USA	500	/			_
		Noblejas	Spain	350	∠	_	_	_
		Peking	China Rep					
		Bata	Guinea	50	∠	_	_	_
		Quito	Ecuador	100	∠	_	_	_
		Riyadh	Saudi Arabia	350	_	_		
		Jerusalem	Israel	300	_		_	_
		Tangier	Morocco	100	/			_
		Santiago	Chile	100		_	_	
		Careysburg	Liberia	250		_	_	
		Carnarvon	Australia	250			_	
11.720	25.60	London	UK	100	_	_	_	
		Limassol	Cyprus	7.5/20	∠	_	_	
		Kinshasa	Zaire	10	∠	_	_	_
		Lvov	Ukraine	240		_		
		Erevan	USSR	100	_	_	_	
		Tula	USSR	50/240			_	_
		Frunze	USSR	100	∠	_		

MHz	Metres	Station	Country	kW	M	M	S	N
		Khabarovsk	USSR	240	/	_	_	
		Sackville	Canada	250	∟	_	_	_
		Peking	China Rep					
		Sottens	Switzerland	500	∟	_	_	_
		Schwarzenburg	Switzerland	100/ 150	/			_
		Leipzig	Germany (E)	100		_	_	
		Okeechobee	USA	100	/		_	_
		Shepparton	Australia	10/100		_	_	_
		Wien	Austria	100	/		_	
		Sofia	Bulgaria	100/500	∟	_	_	_
		Wertachtal	Germany (W)	500	∟	_	_	_
		Lopik	Netherlands	100	/	_	_	_
		Santiago	Chile	100	/	_	_	_
		Riyadh	Saudi Arabia	350	_			
11.725	25.59	S M Galeria	Vatican	100	/	_	_	_
		Delhi	India	20/100	/			
		Aligarh	India	250	_	_	_	_
		Lampertheim	Germany (W)	100		_		
		Biblis	Germany (W)	100	_			
		Lisbonne	Portugal	25/250	∟	_	_	_
		Vologda	USSR	10	_			
		Habana	Cuba	10	∟	_	_	_
		Shepparton	Australia	50	∟	_	_	_
		Malolos	Philippines	100	∟	_	_	_
		Abu Gharib	Iraq	250	_	_	_	_
		Bucuresti	Roumania	250	∟	_	_	_
		Mahe	Seychelles	100	∟	_	_	_
		Tokyo Yamata	Japan	100		_	_	
		Peking	China Rep					
		Lopik	Netherlands	100				
		Voice of Democratic Kampuchea						
		Dacca	Bangladesh	100	/			
11.730	25.58	Lopik	Netherlands	100	∟	_	_	_
		Talata Volon	Malagasy Rep	300/ 600	∟	_	_	_
		Kenga	USSR	100		_	_	
		Khabarovsk	USSR	240	_			
		Tachkent	USSR	50		_	_	
		Vinnitsa	Ukraine	100/500	∟			_
		Greenville	USA	50/500	_	_	_	_
		Allouis	France	100/500	∟	_	_	_
		Noblejas	Spain	350	∟	_	_	_
		Kavalla	Greece	250	∟	_	_	_
		Poro	Philippines	50				
		Tinang	Philippines	250	_	_		
		Agana	Guam	100	/			_
		Quito	Ecuador	100	_	_		_

MHz	Metres	Station	Country	kW	M	M	S	N
		Riyadh	Saudia Arabia	350	L			
		Peking	China Rep					
11.735	25.56	Limassol	Cyprus	20	L	_	_	_
		Peking	Peking					
		Vinnitsa	Ukraine	240		_	_	_
		Kazan	USSR	100	L			_
		Belgrade	Yugoslavia	10/100	L	_	_	_
		Goiana	Brasil	1/7.5	L	_	_	_
		Montevideo	Uruguay	5	L	_	_	_
		Tanger	Morocco	50/100		_	_	
		Delhi	India	100	L	_	_	_
		Allouis	France	100/500		_	_	_
		Kamalabad	Iran	100		_	_	_
		Jerusalem	Israel	100/300	/		_	
		Bucuresti	Roumania	250	L	_	_	
		Sofia	Bulgaria	100	L	_	_	_
		Wavre	Belgium	250	L	_	_	_
		Pori	Finland	100	L			
		Talata Volon	Malagasy Rep	300	L	_	_	
		Fredrikstad	Norway	100		_		
		Sackville	Canada	250				_
		Mt Carlo	Monaco	100	/			_
		Sottens	Switzerland	500				
11.740	25.55	London	UK	250		_		
		Limassol	Cyprus	7.5/20	_	_	_	_
		Masirah	Oman	100	/	_	_	_
		Aligarh	India	250		_	_	_
		Delhi	India	20	/			
		Mexico	Mexico	5	L	_	_	_
		Frunze	USSR	100/500	L	_	_	_
		Novosibirsk	USSR	50	L	_	_	_
		Shepparton	Australia	50	L	_	_	_
		S M Galeria	Vatican	100	L	_	_	
		Cincinnati	USA	250	L		_	_
		Dixon	USA	250	L	_	_	
		Greenville	USA	250		_		
		Ulan Bator	Mongolian Rep	300	_			
		Careysburg	Liberia	250	/	_	_	
		Lopik	Netherlands	100	L	_	_	_
		Kavalla	Greece	250	_	_	_	_
		Kimjae	Korea (S)	100	L	_	_	
		Noblejas	Spain	350	_	_	_	
		Alger	Algeria	100	L	_	_	_
		Talata Volon	Malagasy Rep	300	L	_	_	_
		Mt Carlo	Monaco	100		_	_	_
		Bucuresti	Roumania	250		_	_	
11.745	25.54	Minsk	Bielorussia	240	L	_	_	_
		Alma Ata	USSR	500		_		
		Armavir	USSR	500	/			

MHz	Metres	Station	Country	kW	M	M	S	N
		Allouis	France	100/500	_	_	_	_
		Quito	Ecuador	100	∠	_	_	_
		Godthaab	Greenland	1	∠	_	_	_
		Ekala	Ceylon	35/100	_	_	_	
		Kamalabad	Iran	100				_
		Aligarh	India	250	/			
		Delhi	India	50	_	_	_	_
		Taipei	China					
			Egypt					
11.750	25.53	London	UK	100/250	∠	_	_	_
		Ascension	Ascension	250				_
		Limassol	Cyprus	100				_
		Kranji	Singapore		/			_
		Makassar	Indonesia	1.5	∠	_	_	_
		Tokyo	Japan	1/10	∠	_	_	_
		Kazan	USSR	100			_	_
		Moskva	USSR	100	∠			_
		Lvov	Ukraine	240			_	_
		Caracas	Venezuela	10	∠	_	_	_
		Karachi	Pakistan	50	/			
		Islamabad	Pakistan	250	∠	_	_	_
		Mt Carlo	Monaco	100	∠	_	_	_
		Sofia	Bulgaria	500			_	_
		Agana	Guam	100	/		_	_
		Tinang	Philippines	50				_
		Greenville	USA	250	/			_
11.755	25.52	S Fernando	Argentina	7	∠	_	_	_
		Tripoli	Libya	100	∠	_	_	
		Tbilisi	USSR	240/500	∠		_	_
		Leningrad	USSR	240	∠	_	_	_
		Tchita	USSR	100	_			
		Vladivostock	USSR	100			_	_
		Mahe	Seychelles	100		_	_	_
		Pori	Finland	15/250	∠	_	_	_
		Allouis	France	100			_	_
		Santiago	Chile	100			_	_
		Ekala	Ceylon	35			_	
		Tinang	Philippines					
		Delhi	India			/		
11.760	25.51	London	UK	100		_		
		Limassol	Cyprus	100	∠	_	_	_
		Tinang	Philippines	250	∠	_	_	_
		Erevan	USSR	100		_		
		Krasnoiarsk	USSR	100	_			
		Serpukhov	USSR	100	/			
		Sverdlovsk	USSR	100				_
		Kharkov	Ukraine	100			_	_
		Tanger	Morocco	35/100	∠	_	_	_
		Habana	Cuba	100	∠	_	_	_

MHz	Metres	Station	Country	kW	M	M	S	N
		Manzini	Swaziland	25	_	_	_	_
		Kavalla	Greece	250	∟	_	_	_
		Athinai	Greece	100	∟	_	_	_
		Carnarvon	Australia	250	/	_	_	_
		Shepparton	Australia	100	/	_		
		Jerusalem	Israel	100/300	_	_	_	_
		Santiago	Chile	100	/	_	_	_
		Rarotonga	Cook Is	0.5	=	_		
		Ekala	Ceylon	35	/			_
		Delhi	India	100	/			
11.765	25.50	Kazan	USSR	100	_			
		Leningrad	USSR	500				
		Irkutsk	USSR	100	∟			_
		Bombay	India	100	/			
		Delhi	India	100	∟	_	_	_
		Julich	Germany (W)	100	∟	_	_	_
		Wertachtal	Germany (W)	500	∟	_	_	_
		S Paulo	Brasil	10/25	∟	_	_	_
		Sofia	Bulgaria	50	∟	_	_	_
		Cyclops	Malta	250	∟	_	_	_
		Bocaue	Philippines	50	∟	_	_	_
		Schwarzenburg	Switzerland	250	_	_	_	_
		S M Galeria	Vatican	100	_		_	
		Kigali	Rwanda	250	/	_		
		Santiago	Chile	100	/	_		
		Dacca	Bangladesh					
11.770	25.49	Ascension	Ascension	250	_	_	_	
		Jakarta	Indonesia	20/100	∟	_	_	_
		Ikorodu	Niger	100	/		_	_
		Omsk	USSR	100			_	_
		Kiev	Ukraine	100	/	_	_	_
		Lisbonne	Portugal	100	∟			
		Biblis	Germany (W)	100	/	_	_	
		Lampertheim	Germany (W)	100	_	_	_	
		Playa de Pals	Spain	100/250			_	_
		Tanger	Morocco	25/50			_	_
		Mexico	Mexico	10	∟	_	_	_
		Aligarh	India	250	_	_	_	_
		Delhi	India	100	/	_	_	_
		Kamalabad	Iran	350	_	_		
		Allouis	France	100	∟	_	_	_
		Kabul	Afghanistan	100	/			
		Schwarzenburg	Switzerland	150	/	_	_	
		Okeechobee	USA	100				_
		Alger	Algeria					
11.775	25.48	London	UK	100	/			_
		Antigua	Br W Indies	250	∟	_	_	_
		Schwarzenburg	Switzerland	100/150	_	_		_
		Delhi	India	50	/			

MHz	Metres	Station	Country	kW	M	M	S	N
		Aligarh	India	250				
		Armavir	USSR	240	∟	_	_	_
		Irkutsk	USSR	100/500	∟	_	_	
		Bucuresti	Roumania	18/250	∟	_		_
		Noblejas	Spain	350	∟	_	_	_
		Jerusalem	Israel	20/300	_		_	_
		Carnarvon	Australia	100		_		
		Riyadh	Saudi Arabia	350	/	_	_	
		Santiago	Chile	100		_	_	
		Agana	Guam	100	_			
		Sackville	Canada	50/250	/		_	
11.780	25.47	London	UK	100/250	∟	_	_	_
		Limassol	Cyprus	100	/	_	_	_
		Masirah	Oman	100	∟	_	_	_
		Hurlingham	Argentina	7.5	∟	_	_	_
		Tokyo Yamata	Japan	100	∟	_	_	_
		Lvov	Ukraine	500	_			
		Krasnoiarsk	USSR	100			_	
		Serpukhov	USSR	100		_		
		Tbilisi	USSR	500	/			
		Scituate	USA	100	_			
		Allouis	France	100/500	_	_	_	_
		Brasilia	Brasil	250	∟	_	_	_
		Tinang	Philippines	50	/			_
		Schwarzenburg	Switzerland	100/500	∟	_	_	_
		Kamalabad	Iran	350	_			
		Kimjae	Korea (S)	100		_	_	_
		Lisbonne	Portugal	100	∟			
		Biblis	Germany (W)	100	/			
		Jerusalem	Israel	100/300	∟		_	_
		Agana	Guam	100	_	_		
		Abu Zaabal	Egypt	100			_	
		Ismaning	Germany (W)	100		_		
		Santiago	Chile	100	/			_
		Kavalla	Greece	250		_		
		Mt Carlo	Monaco	100	/			
		Salman Pack	Iraq					
11.782/5	25.46	Schwarzenburg*	Switzerland	30	/			
11.785	25.46	London	UK	100	/		_	
		Montserrat	Br W Indies	50	∟	_	_	_
		Antigua	Br W Indies	250	∟	_	_	_
		Omsk	USSR	100	∟	_	_	_
		Sverdlovsk	USSR	100	∟	_	_	_
		Kigali	Rwanda	250	∟	_	_	_
		K Wusterhausen	Germany (E)	50/100		_	_	
		Wertachtal	Germany (W)	500	∟	_	_	_
		Julich	Germany (W)	100	∟	_	_	_
		Pt Alegre	Brasil	7.5	∟	_	_	_
		Salman Pack	Iraq	100	_	_	_	_

MHz	Metres	Station	Country	kW	M	M	S	N
		Mahe	Seychelles	25/100	∠	−	−	−
		Cyclops	Malta	250	∠	−	−	−
		Abis	Egypt	250			−	
		Arganda	Spain	100	−	−	−	
		Noblejas	Spain	350	/			
		Wavre	Belgium	100/250	−	−	−	−
		Beyrouth	Lebanon	100				−
11.790	25.45	London	UK	250	−	−	−	
		Ascension	Ascension	250	/			−
		Greenville	USA	250	∠	−	−	−
		Cincinnati	USA	175	∠	−	−	−
		Riazan	USSR	240/500	∠	−	−	−
		Nikolaevskamur	USSR	240				−
		Blagovechtchen	USSR	240	∠			
		Shepparton	Australia	10	∠	−	−	−
		Lyndhurst	Australia					
		Bonaire Zuid	Neth Antilles	50	−			
		Bonaire Noord	Neth Antilles	300	∠	−	−	−
		Meyerton	South Africa	100	∠	−	−	−
		Bucuresti	Roumania	18/250	∠	−	−	−
		Hoerby	Sweden	350	−		−	−
		Jakarta	Indonesia	100				
		Wavre	Belgium	100	/			
		Kimjae	Korea (S)	250	∠	−	−	−
		Wien	Austria	100	−	−	−	−
		Abu Zaabal	Egypt	100			−	
		Allouis	France	100/500	∠	−	−	−
		Dacca	Bangladesh	250	/			−
		Bogota	Colombia					
		Mt Carlo	Monaco					
11.795	25.43	Antigua	UK	250	∠	−	−	−
		Tripoli	Libya	100	∠	−	−	
		Julich	Germany (W)	100	∠	−	−	
		Wertachtal	Germany (W)	500	−			
		Kinshasa	Zaire	10	∠	−	−	−
		Kenga	USSR	100			−	
		Moskva	USSR	100	∠		−	
		Tbilisi	USSR	50	−			
		Bogota	Colombia	25	∠	−	−	−
		Cyclops	Malta	250	∠	−	−	−
		K Wusterhausen	Germany (E)	50	−			
		Aligarh	India	250	/			
		Agana	Guam	100		−		
			Egypt					
		Baghdad	Iraq					
11.800	25.42	Ejura	Ghana	250	∠	−	−	−
		Ekala	Ceylon	100	∠	−	−	−
		Roma	Italy	60/100	∠	−	−	−
		Kiev	Ukraine	240/500	∠			−

MHz	Metres	Station	Country	kW	M	M	S	N
		Mt Carlo	Monaco	100	—			
		Santiago	Chile	100				
		Malolos	Philippines	50/100	—			
		Bonaire Zuid	Neth Antilles	50		—	—	—
		Meyerton	South Africa	250/500	—	—	—	—
		Mahe	Seychelles	25/100	—		—	
		Quito	Ecuador	100	∟	—		
		Ankara	Turkey	250		—	—	
		Warszawa	Poland	1	∟	—	—	—
		S Gabriel	Portugal	100	—	—	—	—
		Wellington	New Zealand	7.5	—	—	—	
		Carnarvon	Australia	250	∟	—		
		Pori	Finland	250	—	—	—	
		Hoerby	Sweden	350	—	—		
		Riyadh	Saudi Arabia	350				—
		Wavre	Belgium					
		Alger	Algeria					
		Okeechobee	USA					
11.805	25.41	London	UK	100/250	—		—	—
		Dixon	USA	100/250	∟	—	—	—
		Greenville	USA	50/500	—	—	—	
		Okeechobee	USA	100				—
		Scituate	USA	50/100			—	—
		Kazan	USSR	50	∟	—	—	—
		Tbilisi	USSR	240/500		—	—	
		Malolos	Philippines	100	∟	—	—	—
		Tinang	Philippines	250	∟	—	—	—
		Kavalla	Greece	250	∟			
		Rio de Janeiro	Brasil	10	∟	—	—	—
		Careysburg	Liberia	50	∟	—	—	—
		Tanger	Morocco	10	/			
		Allouis	France	100/500	∟	—	—	—
		Mahe	Seychelles	25/100	/	—	—	—
		Pt au Prince	Haiti	100				—
		Kabul	Afghanistan					
		Nauen	Germany (E)	100	/			—
		Ismaning	Germany (W)	100	/			
11.810	25.40	Antigua	Br W Indies	125/250	∟	—	—	—
		Alger	Algeria	100	∟	—	—	—
		Aligarh	India	100/250	—	—	—	—
		Delhi	India	100	/			
		Shepparton	Australia	50/100	—	—	—	—
		Simferopol	Ukraine	240		—	—	
		Serpukhov	USSR	100/240	∟			—
		Tchita	USSR	100	∟			
		Komsomolskamur	USSR	100			—	
		Julich	Germany (W)	100	∟	—	—	—

MHz	Metres	Station	Country	kW	M	M	S	N
		Noblejas	Spain	350	∠	—	—	—
		Roma	Italy	60/100	∠	—	—	—
		Las Mesas	Canary Is	50	—	—		
		Bucuresti	Roumania	18/120	∠	—		—
		Jerusalem	Israel	50				—
		Agana	Guam	100	—	—	—	—
		Kavalla	Greece	250		—		
		Cyclops	Malta	250	/	—	—	—
		S M Galeria	Vatican	100	∠	—	—	—
		K Wusterhausen	Germany (E)	100	/			—
		Hoerby	Sweden	350	/			—
		Manzini	Swaziland	25				—
		Tinang	Philippines	250	/			—
11.815	25.39	Warszawa	Poland	100	∠	—	—	—
		Bonaire Zuid	Neth Antilles	50/250	—	—	—	—
		Goiana	Brasil	7.5	∠	—	—	—
		Biblis	Germany (W)	100	∠	—	—	—
		Lisbonne	Portugal	50	∠	—	—	—
		Tokyo Yamata	Japan	100	∠	—	—	—
		Scituate	USA	50/100				
		Okeechobee	USA	100	—			
		Tinang	Philippines	250	∠			
		Aligarh	India	250	∠	—	—	—
		Noblejas	Spain	350	∠	—	—	—
		Las Mesas	Canary Is	50	/	—	—	—
		Agana	Guam	100				
11.820	25.38	Antigua	Br W Indies	125/200		—		
		Limassol	Cyprus	100	—	—		
		Ascension	Ascension	125/250	∠			
		Maputo	Mozambique	120	∠		—	—
		Krasnoirask	USSR	100		—		
		Frunze	USSR	240/500	∠			
		Khabarovsk	USSR	100				
		Tbilisi	USSR	240	/			—
		Voronej	USSR	100		—	—	
		Shepparton	Australia	50/100	/	—	—	—
		Bocaue	Philippines	50	—	—		
		Delano	USA	250	∠		—	—
		S M Galeria	Vatican	100				
		Wertachtal	Germany (W)	500	∠	—	—	—
		Quito	Ecuador	100	—			
		Sofia	Bulgaria	100	—			—
		Riyadh	Saudi Arabia	350		—		
		Kabul	Afghanistan	100				—
11.825	25.37	Bogota	Colombia	25	∠	—	—	—
		Shepparton	Australia	10/100	∠	—	—	—
		Papeete	Tahiti	20	∠	—	—	—
		Holzkirchen	Germany (W)	10	∠	—	—	—

MHz	Metres	Station	Country	kW	M	M	S	N
		Lisbonne	Portugal	50	∠	_	_	_
		Sines	Portugal	250		_		
		Recife	Brasil	10/15	∠	_	_	_
		Taipei	China Nat	25				
		Sackville	Canada	50/250	∠		_	
		Delhi	India	50		_	_	_
		Duchanbe	USSR	100	∠			
		Allouis	France	100		_	_	_
		Beyrouth	Lebanon	100	_			
		Malolos	Philippines	100				
		Pori	Finland	250			_	_
		Wien	Austria	100	/		_	_
		Nauen	Germany (E)	50	/			_
		Abu Ghraib	Iraq					
11.830	25.36	London	UK	100/250		_		_
		Bonaire Zuid	Neth Antilles	50/260		_	_	
		Greenville	USA	500	_			_
		Delano	USA	250			_	_
		Cincinnati	USA	250	∠		_	_
		Dixon	USA	250	∠	_	_	_
		Bombay	India	100	∠	_	_	_
		Moskva	USSR	240	∠	_	_	_
		Malolos	Philippines	100		_	_	_
		Tinang	Philippines	250				_
		Kavalla	Greece	250	∠	_	_	_
		Bucuresti	Roumania	18/250	∠	_	_	_
		Arganda	Spain	100	∠	_	_	_
		S M Galeria	Vatican	100	∠	_	_	_
		Quito	Ecuador	100				_
		V of Malayan Revln						
		Monrovia	Liberia	50	∠	_	_	_
11.835	25.35	London	UK	100				_
		Ekala	Ceylon	10/35	∠	_	_	_
		Omdurman	Sudan	120		_	_	
		Serpukhov	USSR	100		_	_	_
		Krasnoiarsk	USSR	200	∠	_		_
		Tachkent	USSR		/			_
		Montevideo	Uruguay	5	∠	_	_	_
		Mt Carlo	Monaco	100	∠	_	_	
		Quito	Ecuador	100	∠	_	_	
		Monrovia	Liberia	50	∠	_	_	_
		Poro	Philippines	35/100	/	_	_	_
		Greenville	USA	250	/	_	_	_
		Cincinnati	USA	175		_		
		Shepparton	Australia	10		_		_
		Carnarvon	Australia	100	/		_	
		Cap Haitien	Haiti					
		Franceville	Gabon Rep	500			_	

MHz	Metres	Station	Country	kW	M	M	S	N
		Delhi	India	100	/			
		Pori	Finland	250	/			
11.840	25.34	Nauen	Germany (E)	100			—	—
		S Gabriel	Portugal	100	∠	—	—	—
		Warszawa	Poland	40/100	∠	—	—	—
		Kavalla	Greece	250	∠	—	—	—
		Greenville	USA	50/250	∠	—	—	—
		Tokyo Yamata	Japan	50/100	∠	—	—	—
		Poro	Philippines	35	—			—
		Bucuresti	Roumania	50/120	/	—	—	—
		Delhi	India	20	∠	—	—	—
		Quito	Ecuador	50	∠			
		Noblejas	Spain	350	∠	—	—	
		Careysburg	Liberia	50/250			—	
		Biblis	Germany (W)	100				—
		Kimjae	Korea (S)	100			—	—
		Mahe	Seychelles	100	—	—	—	
		Allouis	France	100	/		—	—
		Wellington	New Zealand	7.5			—	—
		Jerusalem	Israel					
		Agana	Guam	100	/			
11.845	25.33	London	UK	100/250				
		Moskva	USSR	250				—
		Serpukhov	USSR	100	—			
		Kazan	USSR	100	∠	—	—	—
		Montevideo	Uruguay	10	∠	—	—	—
		Allouis	France	100/500	∠	—	—	—
		S M Galeria	Vatican	100	∠	—	—	—
		Lopik	Netherlands	100	∠	—	—	—
		Tirane	Albania					
		Sackville	Canada	50/250	∠	—	—	—
		Hoerby	Sweden	350	∠	—	—	—
		Kimjae	Korea (S)	250	—			
		Riyadh	Saudi Arabia	350		—		
		Greenville	USA	250			—	
		Okeechobee	USA	100				
		Jerusalem	Israel	300	/			
		Delhi	India	100	/			
			Philippines	250			—	
		Athinai	Greece					
11.850	25.32	London	UK	100/250	—	—	—	—
		Kranji	Singapore	100	/			
		Limassol	Cyprus	100				
		Delano	USA	250	/	—		
		Scituate	USA	100	—			
		Okeechobee	USA	100				
		Delano	USA	250				—
		Dixon	USA	100/200	—		—	—
		Asuncion	Paraguay	5	/	—	—	—

MHz	Metres	Station	Country	kW	M	M	S	N
		Delhi	India	50/100	/	_	_	_
		Ejura	Ghana	250	∟	_	_	_
		Fredrikstad	Norway	100/250	∟	_	_	_
		Schwarzenburg	Switzerland	150	∟			_
		Vinnitsa	Ukraine	240		_	_	
		Konevo	USSR	240	∟	_	_	_
		Frunze	USSR	240	∟	_	_	
		Kazan	USSR	100	∟	_	_	
		Julich	Germany (W)	100	∟	_	_	_
		Wertachtal	Germany (W)	500	∟	_	_	_
		Kavalla	Greece	250	∟	_	_	
		Noblejas	Spain	350	_	_	_	
		Wavre	Belgium	250	∟	_	_	_
		Nauen	Germany (E)	50	_			_
		Wellington	New Zealand	7.5	_		_	
		Karlsborg	Sweden	350	_			
		Hoerby	Sweden	350	_			
		Agana	Guam	100			_	_
		Sofia	Bulgaria	500		_	_	
		Allouis	France	500	/		_	_
		Kimjae	Korea (S)	250			_	_
11.855	25.31	London	UK	100/250	/		_	_
		Bocaue	Philippines	50	∟		_	_
		Scituate	USA	50/100		_	_	
		Okeechobee	USA	100	∟	_	_	_
		Greenville	USA	100	_			
		Delhi	India	20/100	∟	_	_	_
		Jeddah	Saudi Arabia	50	∟	_	_	_
		Riyadh	Saudi Arabia	350		_	_	_
		Ulan Bator	Mongolian Rep					
		Sackville	Canada	50/250		_	_	_
		Praha	Czechoslovakia	120	∟	_	_	_
		Biblis	Germany (W)	50	_			
		Lampertheim	Germany (W)	50				_
		Lisbonne	Portugal	50/250		_	_	_
		Kavalla	Greece	250		_		
		Arganda	Spain	100	_			
		Santiago	Chile	100	_	_		
		Allouis	France	100/500			_	_
		Shepparton	Australia	50/100	∟	_	_	_
		Carnarvon	Australia	100			_	_
		Mt Carlo	Monaco	100	_	_		
		Tokyo Yamata	Japan	100	∟	_	_	
		Hoerby	Sweden	350	_			
		Mahe	Seychelles	50	/	_	_	_
		S M Galeria	Vatican	100				
11.860	25.30	Ascension	Ascension	125/250	∟	_	_	_
		Gorkii	USSR	240	∟	_	_	_
		Krasnoiarsk	USSR	100	_	_	_	

MHz	Metres	Station	Country	kW	M	M	S	N
		Iujnsakhalinsk	USSR	100	_			_
		Montevideo	Uruguay	10	L	_	_	_
		Taipei	China Nat	50				
		Fredrikstad	Norway	100/250	L	_	_	
		Kimjae	Korea (S)	250	L	_		
		Allouis	France	100	/	_	_	_
		Monrovia	Liberia	50	L	_	_	_
		Bucuresti	Roumania	250	_			
		Mahe	Seychelles	100	L	_	_	
		Sofia	Bulgaria	500	/			
		Sackville	Canada	250				
11.865	25.28	London	UK	100/250	L	_	_	_
		Antigua	Br W Indies	250	L	_	_	_
		Limassol	Cyprus	100	/			_
		Kranji	Singapore	250	/		_	_
		Recife	Brasil	1/7.5	L	_	_	_
		Lubumbashi	Zaire	100	L	_	_	_
		Tirane	Albania					
		Jayapura	Indonesia	25	L	_	_	_
		Dixon	USA	100/250	_		_	_
		Delano	USA	100	/	_		
		Julich	Germany (W)	100	L	_		
		Wertachtal	Germany (W)	500	L	_	_	
		Sines	Portugal	250	L	_	_	
		Mahe	Seychelles	30/100	L	_	_	
		Cyclops	Malta	250	L	_	_	
		Habana	Cuba	10/100	_	_	_	
		Lampertheim	Germany (W)	100	_			
		Novosibirsk	USSR	100			_	
		Allouis	France	500		_	_	
		Kamalabad	Iran	100	_			
		Agana	Guam	100	_			
		Carnarvon	Australia	100				_
		Riyadh	Saudi Arabia	350				_
		Quito	Ecuador	100				
11.870	25.27	Khabarovsk	USSR	240			_	_
		Serpukhov	USSR	240	L	_	_	_
		Kaunas	USSR	240	/	_		
		Kenga	USSR	240	L			_
		Bombay	India	100	L	_	_	
		Schwarzenburg	Switzerland	250	L	_	_	
		Wien	Austria	100	_	_	_	_
		Fredrikstad	Norway	100/250				_
		Riyadh	Saudi Arabia	350	L	_	_	_
		Malolos	Philippines	100	L	_	_	_
		Sofia	Bulgaria	50/500	L	_	_	_
		Carnarvon	Australia	100				
		Lyndhurst	Australia	35				
		Ekala	Ceylon	100	/		_	_

MHz	Metres	Station	Country	kW	M	M	S	N
		Monrovia	Liberia					
11.875	25.26	Malolos	Philippines	100	∠	_	_	_
		Salvador	Brasil	10	∠	_	_	_
		Tokyo Yamata	Japan	100	∠	_	_	_
		Biblis	Germany (W)	100	∠			
		Lampertheim	Germany (W)	100	/			
		Holzkirchen	Germany (W)	10		_	_	_
		Playa de Pals	Spain	50/500	∠	_	_	_
		Lisbonne	Portugal	250	/			
		S Gabriel	Portugal	100	∠	_	_	_
		Okeechobee	USA	100		_		
		K Wusterhausen	Germany (E)	100	∠			
		S M Galeria	Vatican	100		_	_	_
		Kavalla	Greece	250	∠		_	_
		Managua	Nicaragua	100	∠	_	_	_
		Jerusalem	Israel	45/300		_		_
		Roma	Italy	100	/	_	_	_
		Delhi	India	100	/			
11.880	25.25	Lusaka	Zambia	20/50	∠	_	_	_
		Lomas Mirador	Argentina	20	∠	_	_	_
		Lyndhurst	Australia	10	∠	_	_	_
		Shepparton	Australia	50	∠	_	_	_
		Mexico	Mexico	5	∠	_	_	_
		Armavir	USSR	240			_	_
		Moskva	USSR	240	∠		_	
		Tchita	USSR	240			_	
		Ivanofrankovsk	Ukraine	100/240	∠			_
		Delhi	India	100				
		Noblejas	Spain	350	∠	_	_	_
		Las Mesas	Canary Is	50		_	_	_
		Riyadh	Saudi Arabia	350		_		
		Monrovia	Liberia	50			_	_
		Wavre	Belgium					
11.885	25.24	Novosibirsk	USSR	50	/			_
		Bucuresti	Roumania	25/120	∠	_	_	_
		Karachi	Pakistan	50/120	∠	_	_	_
		Islamabad	Pakistan	100		_		_
		Montevideo	Uruguay	10	∠	_	_	_
		Lisbonne	Portugal	250		_		_
		Biblis	Germany (W)	100	_			
		Holzkirchen	Germany (W)	10		_		
		Lampertheim	Germany (W)	100	∠	_	_	_
		Playa de Pals	Spain	100	/	_	_	
		Rio de Janeiro	Brasil	10	∠	_	_	_
		Okeechobee	USA	100				_
		Meyerton	South Africa	100	∠	_	_	_
		Pyongyang	Korea (N)					
		Delhi	India	100				_
		Riyadh	Saudi Arabia					

MHz	Metres	Station	Country	kW	M	M	S	N
11.890	25.23	K Wusterhausen	Germany (E)	100				
		Nauen	Germany (E)	100	L	_	_	
		Bocaue	Philippines	50	L	_	_	_
		Greenville	USA	250/500	L	_	_	_
		Kenga	USSR	100	_			
		Frunze	USSR	240	L	_		
		Erevan	USSR	500				_
		Riazan	USSR	240				
		Tchita	USSR	240	/			
		Allouis	France	100/500	_	_	_	
		Riyadh	Saudi Arabia	350	/		_	_
		Seeb	Oman	50	L	_	_	
		Dacca	Bangladesh	100	L	_	_	_
		Santiago	Chile	100	/	_	_	_
		Agana	Guam	100	_			
		Noblejas	Spain	350	/	_	_	
		Kabul	Afghanistan					
11.895	25.22	Delhi	India	20/100	_	_	_	
		Lisbonne	Portugal	50/250	L	_	_	_
		Biblis	Germany (W)	100	/			_
		Lampertheim	Germany (W)	100	_	_		
		Playa de Pals	Spain	250		_	_	
		Cincinnati	USA	250	L	_	_	_
		Fredrikstad	Norway	250	_	_		_
		Jigulevsk	USSR	240	/			_
		Dakar	Senegal	100	/	_	_	
		Wien	Austria	100	_	_	_	_
		Bata	Guinea	50	L	_	_	
		Dacca	Bangladesh	100	_			
		Carnarvon	Australia	250		_	_	
11.900	25.21	Armavir	USSR	100	_	_		
		Komsomolskamur	USSR	100	L			
		Duchanbe	USSR	240		_	_	
		Lvov	Ukraine	240/500	L			_
		Meyerton	South Africa	250/500	L	_	_	_
		Kajang	Malaysia	100	L	_	_	
		Montevideo	Uruguay	20	L	_	_	_
		Greenville	USA	50/250	L	_	_	
		Dixon	USA	250				_
		Delano	USA	250	/			
		Ikorodu	Niger	100	L	_	_	_
		Dacca	Bangladesh	100/250				_
		Bonaire Zuid	Neth Antilles	250	L	_	_	_
		Agana	Guam	100	_	_	_	
		Quito	Ecuador	100	L	_	_	_
		Riyadh	Saudi Arabia	350	_	_	_	
11.905	25.20	London	UK	100/250	L	_	_	_
		Taipei	China Rep	3				

MHz	Metres	Station	Country	kW	M	M	S	N
		Julich	Germany (W)	100	∠	—	—	—
		Wertachtal	Germany (W)	500	∠	—	—	—
		Roma	Italy	60/100	∠	—	—	—
		Frunze	USSR	100	/	—		—
		Greenville	USA	50/250	∠	—	—	—
		Red Lion	USA	50	—			
		Sines	Portugal	250	∠	—	—	—
		Patumthani	Thailand	100	∠	—	—	—
		Hoerby	Sweden	350	/	—	—	—
		Karlsborg	Sweden	350	/			
		Babel	Iraq	500	—	—		
		Quito	Ecuador	100				
		Sackville	Canada	250				—
		Riyadh	Saudia Arabia	350	/			
11.910	25.19	London	UK	100/250		—	—	—
		Montserrat	Br W Indies	15/50				
		Masirah	Oman	100/200	∠	—	—	—
		Kranji	Singapore	100	/			—
		Jaszbereny	Hungary	250	∠	—	—	—
		Diosd	Hungary	100	∠	—	—	—
		Moskva	USSR	240	—			
		Sverdlovsk	USSR	100	/	—		—
		Komsomolskamur	USSR	100	∠			
		Quito	Ecuador	100	∠	—	—	—
		Kimjae	Korea (S)	10	/			
		Noblejas	Spain	350	—	—	—	—
		Riyadh	Saudi Arabia	350	/	—	—	—
		Pori	Finland	250				—
		Manzini	Swaziland					
		Lampertheim	Germany (W)	100	/			
		Islamabad	Pakistan	250	/			
		Allouis	France	100/500	/			
11.915	25.18	London	UK	250	/	—		
		Abu Zaabal	Egypt	100		—		
		Concepcion	Paraguay	100	∠	—	—	—
		Petropavlo Kam	USSR	50/100		—	—	
		Pt Alegre	Brasil	7.5	∠	—	—	—
		Quito	Ecuador	50/100	∠	—	—	—
		Delano	USA	200	—	—		
		Greenville	USA	250	—			
		Tanger	Morocco	35/100	∠		—	—
		Careysburg	Liberia	250	/		—	—
		Sines	Portugal	250	∠	—	—	—
		Playa de Pals	Spain	100		—		
		Holzkirchen	Germany (W)	10		—		
		Taipei	China Nat					
		Delhi	India	100	—	—	—	—
		Islamabad	Pakistan	250	/	—	—	—
		Karachi	Pakistan	50			—	

MHz	Metres	Station	Country	kW	M	M	S	N
		Hoerby	Sweden	350	_			
		Riyadh	Saudi Arabia	350	/	_	_	
		Sackville	Canada	250		_		
		Tirana	Albania					
		Munich	Germany (W)	100				
11.920	25.17	Kranji	Singapore	100	/			
		Dixon	USA	100				_
		Greenville	USA		_	_	_	_
		Bocaue	Philippines	50	∠	_	_	_
		Iba	Philippines	100				_
		Arganda	Spain	100	/			
		Noblejas	Spain	350				
		Moskva	USSR	240	∠	_		_
		Novosibirsk	USSR	100		_	_	
		Baku	USSR	240	∠	_	_	_
		Kazan	USSR	100	∠			
		Alma Ata	USSR	100	_			
		Abidjan	Ivory Coast	100	∠	_	_	_
		Cyclops	Malta	250	_			
		Ismaning	Ismaning	100	∠			
		Nauen	Germany (E)	100	∠		_	
			China Rep					
		Amman	Jordan	100	∠	_	_	_
		Tirana	Albania					
		Sofia	Bulgaria					
11.925	25.16	London	UK	100/250	/	_	_	
		Kranji	Singapore	100	/			
		S Gabriel	Portugal	50	∠	_	_	_
		Lisbonne	Portugal	50		_	_	
		Lampertheim	Germany (W)	100	∠			
		Biblis	Germany (W)	100			_	
		Playa de Pals	Spain	250				_
		Arganda	Spain	100	_		_	_
		Noblejas	Spain	350	∠	_	_	_
		S Paulo	Brasil	10	∠	_	_	_
		Krasnoiarsk	USSR	240	_			
		Tachkent	USSR	50			_	
		Tchita	USSR	500	/		_	
		Kiev	Ukraine	100	∠		_	
		Kavalla	Greece	250	∠	_		
		Bonaire Zuid	Neth Antilles	250	∠	_	_	_
		Ikorodu	Niger	100	_	_		
		Santiago	Chile	100	_			
		Tinang	Philippines	250		_		_
		Alger	Algeria	100	∠	_	_	_
		Allouis	France	100		_	_	
		Babel	Iraq	500			_	_
		Beyrouth	Lebanon	100			_	
		Okeechobee	USA	100				_

MHz	Metres	Station	Country	kW	M	M	S	N
		Scituate	USA	50				—
11.930	25.15	Habana	Cuba	50	∠	—	—	—
		Achkhabad	USSR	240/500		—	—	
		Moskva	USSR	240	∠	—	—	
		Armavir	USSR	120	∠	—	—	—
		Tinang	Philippines	50/250	∠	—	—	—
		Poro	Philippines		/			—
		Lopik	Netherlands	100	∠	—	—	—
		Kamalabad	Iran	100		—	—	
		Noblejas	Spain	350	∠	—	—	—
		Arganda	Spain	100	/			—
		Allouis	France	100/500	∠	—	—	—
		Monrovia	Liberia	50	/			
11.935	25.14	London	UK	100/250	∠	—	—	—
		Wertachtal	Germany (W)	500	/	—	—	—
		Biblis	Germany (W)	100		—		
		Lampertheim	Germany (W)	100			—	
		Playa de Pals	Spain	250	∠	—		
		S Gabriel	Portugal	100	∠	—	—	—
		Meyerton	South Africa	100	∠	—	—	—
		Delhi	India	20	∠	—	—	
		Greenville	USA	50/250				—
		Curitiba	Brasil	7.5/25	∠	—	—	—
		Tirane	Albania					
		Carnarvon	Australia	250	—	—	—	—
		Shepparton	Australia	100	—			
		Fredrikstad	Norway	100/250		—		—
		Ekala	Ceylon	35	∠	—	—	—
		Tula	USSR	120	—		—	
		Mt Carlo	Monaco	100	—	—	—	
		Babel	Iraq	500				
		Lopik	Netherlands	100			—	—
		Hoerby	Sweden	350	X		—	
11.940	25.13	Bucuresti	Roumania	120/250	∠	—	—	—
		Sulaibiyah	Kuwait	250	∠	—	—	—
		Blagovechtchen	USSR	240		—	—	
		Krasnoiarsk	USSR	240	—		—	
		Sverdlovsk	USSR	100	∠			
		Singapore	Singapore	50	∠	—	—	—
		Tokyo Yamata	Japan	50	∠	—	—	—
		Encarnacion	Paraguay	5	∠	—	—	—
		Sackville	Canada	50	—	—	—	—
		Wavre	Belgium	100/250	∠	—	—	—
		Ekala	Ceylon	35	—		—	
11.945	25.12	London	UK	100/250	∠	—	—	—
		Encarnacion	Paraguay	5	∠	—	—	—
		Wertachtal	Germany (W)	500	∠	—	—	—
		Peking	China Rep					
		Monrovia	Liberia	50	—	—	—	—

136

MHz	Metres	Station	Country	kW	M	M	S	N
		Delhi	India	100	_	_	_	_
		Kimjae	Korea (S)	100	L	_	_	_
		Mahe	Seychelles	100	L	_	_	_
		Sackville	Canada	50/250	L	_	_	_
		Noblejas	Spain	350/700	L	_	_	_
		Playa de Pals	Spain	250				_
		Wellington	New Zealand	7.5				_
		Allouis	France	100				
		Quito	Ecuador	100				
11.950	25.11	Diriyya	Saudi Arabia	50	L	_	_	_
		Alma Ata	USSR	100	L	_	_	_
		Tula	Ukraine	100		_	_	
		Kharkov	Ukraine	240	L	_	_	
		Rio de Janeiro	Brasil	7.5/10	L	_	_	_
		Lopik	Netherlands	100	L	_	_	_
		Malolos	Philippines	50		_	_	_
		Tokyo Yamata	Japan	200		_	_	_
		Bucuresti	Roumania	120/250	L	_	_	_
		Bonaire Zuid	Neth Antilles	50	/		_	_
		Tirane	Albania					
11.9525	25.10	Varberg*	Sweden	100				_
11.955	25.09	London	UK	250	L	_	_	_
		Kranji	Singapore	125/250	/	_	_	_
		Limassol	Cyprus	100	/	_	_	
		Masirah	Oman	100	_			
		Ivanofrankovsk	Ukraine	240		_	_	
		Montevideo	Uruguay	10	L	_	_	_
		Greenville	USA	250/500	L	_	_	_
		Tanger	Morocco	35/100	_		_	
		Hoerby	Sweden	350	/	_	_	_
		Malolos	Philippines					
		Ekala	Ceylon	10/100		_		
		Manzini	Swaziland	25	_	_		_
		Jerusalem	Israel	300		_		
		Lopik	Netherlands	100	_	_	_	
		Athinai	Greece	100				
		Allouis	France	100/500	_		_	
		Luanda	Angola	100	/		_	_
		K Wusterhausen	Germany (E)	100		_		
		Mt Carlo	Monaco	100			_	
		Ankara	Turkey	250	/			_
		Roma	Italy	60				_
		Meyerton	South Africa	250				_
		Islamabad	Pakistan	250				_
11.960	25.08	London	UK	100/250	L	_	_	_
		Bamako	Malawi	50		_	_	
		Armavir	USSR	240	L	_		
		Serpukhov	USSR	100	/		_	
		Tachkent	USSR	100/500	L	_	_	

MHz	Metres	Station	Country	kW	M	M	S	N
		Tbilisi	USSR	240				_
		Kenga	USSR	240		_	_	
		Krasnoiarsk	USSR	100	∟	_	_	_
		Kazan	USSR	100	/			
		Tokyo Yamata	Japan	200	/			_
		Jerusalem	Israel	300	∟	_	_	_
		Tanger	Morocco	35	_			
		Quito	Ecuador	100	∟	_	_	_
		Wellington	New Zealand	7.5	∟	_		_
		Sackville	Canada	250			_	
		Greenville	USA	50/250	∟		_	_
		Leipzig	Germany (E)	100	/			_
11.965	25.07	Kigali	Rwanda	250	∟	_	_	_
		Allouis	France	100	_	_	_	_
		Kazan	USSR	240		_	_	
		Tinang	Philippines	250	∟	_	_	_
		Poro	Philippines	100	_	_		
		S Paulo	Brasil	7.5	∟	_	_	_
		Conakry	Guinea	100	∟	_	_	_
		Noblejas	Spain	350		_	_	_
		Greenville	USA	50	∟			
		Riyadh	Saudi Arabia	350	_			
		Beyrouth	Lebanon	100		_		
		Kimjae	Korea (S)	250	/		_	_
		Tirane	Albania					
		Franceville	Gabon Rep	500			_	
		Ankara	Turkey	250	/			_
		Delhi	India	20	/			
		Tangier	Morocco	50				
11.970	25.06	Montserrat	Br W Indies	15	_	_	_	
		Sofia	Bulgaria	50	/	_	_	_
		Lisbonne	Portugal	50/100		_	_	_
		Biblis	Germany (W)	100		_	_	
		Lampertheim	Germany (W)	250	∟	_	_	
		Playa de Pals	Spain	100/250	∟			_
		Sfax	Tunisia	50/100				
		Habana	Cuba	10/50	∟	_	_	_
		Nauen	Germany (E)	50		_	_	
		K Wusterhausen	Germany (E)	100	∟	_	_	_
		Komsomolskamur	USSR	100		_	_	
		Frunze	USSR	240/500	∟			_
		Kharkov	Ukraine	240		_	_	
		Bucuresti	Roumania	120/250	∟	_	_	_
		Malolos	Philippines	100	∟	_	_	_
		Kimjae	Korea (S)	100	∟	_	_	_
		Kathmandu	Nepal	100	∟		_	_
		Islamabad	Pakistan	10	/			
11.975	25.05	Berlin	Germany (E)					
		Moskva	USSR					

MHz	Metres	Station	Country	kW
11.980	25.04	Bucurest	Roumania	
		Peking	China Rep	
		Moskva	USSR	
		V of Democratic Kampuchea		
11.985	25.03	Orcha	USSR	60
		Tirane	Albania	240
		Kabul	Afghanistan	
11.990	25.02	Prague	Czechoslovakia	100
			USSR	50
		Peking	China Rep	
		Sulaibiyah	Kuwait	
		Warszawa	Poland	
		V of Democratic Kampuchea		
11.995	25.01	Moskva	USSR	
12.000	25.00		USSR	100
12.005	24.99	Cairo	Egypt	100
		Minsk	USSR	
12.006	24.99	V NUFK		
12.010	24.98		USSR	
		Monrovia*	Liberia	
12.015	24.97	Wien	Austria	
		Peking	China Rep	
12.020	24.96		USSR	
		Monrovia*	Liberia	
12.025	24.95		USSR	
12.030	24.94		USSR	100
12.033	24.93	Hanoi	Vietnam	
12.035	24.93		USSR	
12.040	24.92	London	UK	
		Yerevan	USSR	100
12.045	24.91	Moskva	USSR	
		Cairo	Egypt	
12.050	24.90		USSR	
		Cairo	Egypt	
12.055	24.89	Peking	China Rep	
		Moskva	USSR	
12.060	24.88		USSR	
		Peking	China Rep	
12.070	24.86	Kiev	Ukraine	100
		Moskva*	USSR	
		Ulan Bator	Mongolian Rep	
		Cairo	Egypt	
12.075	24.84		USSR	
12.077	24.84	Jerusalem	Israel	
12.080	24.83	Peking	China Rep	
12.085	24.82		Kuwait	
12.095	24.80	London	UK	

MHz	Metres	Station	Country	kW
12.100	24.79		USSR*	
12.110	24.77	Peking	China	
12.113	24.76	Riyadh*	Saudi Arabia	
12.120	24.75	Peking	China Rep	
12.127	24.74	London*	UK	
12.140	24.71		USSR	
12.165	24.66		USSR*	
12.175	24.64		USSR*	
		Reykjavik*	Iceland	
12.179	24.63	London*	UK	
12.190	24.61		Australia	
12.200	24.59	Peking	China Rep	
12.205	24.58		USSR*	
12.2225	24.54		USSR*	
12.240	24.51	Magadan*	USSR	50
12.246	24.50	Liberty*		
12.250	24.49		USSR*	
12.25275	24.48		USSR*	
12.280	24.43		USSR	
12.290	24.41		Australia	
			Germany (W)*	
12.295	24.40	Mt Carlo*	Monaco	
12.420	24.15	Peking	China Rep	
12.450	24.10	Peking	China Rep	120
12.460	24.08	Paris	France	
13.272	22.60	New York	USA	
13.370	22.44		USSR*	
13.380	22.42		USSR*	
13.396	22.39	Riyadh	Saudi Arabia	
13.491	22.24	Greenville*	USA	
13.512	22.20	Julich/Wertachtal*	Germany (W)	
13.590	22.08		USSR*	
13.690	21.91	RFE*		
13.710	21.88		USSR*	
13.760	21.80		USSR*	
13.820	21.71		USSR*	
13.849	21.66	Bogota	Colombia	
13.860	21.65	Delano*	USA	
13.960	21.49		USSR*	
14.270	21.02	Tirane	Albania	
14.290	20.99		USSR	
14.440	20.78	R Free Portugal		
			USSR	
14.595	20.55		USSR	
14.670	20.45	'CHU' Time Sig	Canada	
14.700	20.41			
14.712	20.39	RFE/R Liberty*		
14.715	20.39	RFE/R Liberty*		
14.820	20.24	Peking	China Rep	

MHz	Metres	Station	Country	kW	M	M	S	N
14.850	20.20		USSR*					
14.990	20.01	Alger	Algeria					
14.996	20.01	Freq Standard	USSR					
15.000	20.00	Boulder Freq Std	USA					
		Honolulu Freq Std	Hawaii					
		Tokyo Freq Std	Japan					
15.009	19.99	Hanoi	Vietnam					
15.020	19.97	Peking	China Rep					
		Hanoi	Vietnam					
15.030	19.96	Peking	China Rep	120				
15.040	19.95	Peking*	China Rep					
15.045	19.94	Peking	China Rep	240				
			USSR					
15.050	19.93	Peking	China Rep					
		Grenada	Br W Indies					
15.060	19.92	Peking	China Rep	240				
			USSR					
		Riyadh	Saudi Arabia					
15.070	19.91	London	UK					
		Peking	China Rep					
		Conakry	Guinea					
		Jerusalem	Israel					
15.080	19.89	Delhi	India	100				
		Bombay	India					
		Peking	China Rep					
15.084	19.89	Teheran	Iran	250				
15.090	19.88	Monrovia	Liberia					
15.095	19.87	Peking	China Rep	240				
15.100	19.87	Jerusalem	Israel					
		Moskva	USSR					
		Berlin	Germany (E)					
		Tripoli	Libya					
15.105	19.86	London	UK	100	/		—	—
		Limassol	Cyprus	100	—	—		—
		Ascension	Ascension	250	∠	—	—	—
		Jerusalem	Israel	100/300	∠	—	—	—
		Tokyo Yamata	Japan	100	∠	—	—	—
		K Wusterhausen	Germany (E)	50/100	∠			
		Rio de Janeiro	Brasil	7.5	∠			—
		Wien	Austria	100	—	—	—	—
		Pori	Finland	100	—	—	—	—
		Cyclops	Malta	250	/	—	—	
		Julich	Germany (W)	100	/	—	—	
		Wertachtal	Germany (W)	500	/		—	—
		Shepparton	Australia	100				—
		Grenada	Br W Indies	10				
15.110	19.85	London	UK					
		Alma Ata	USSR	240				—
		Krasnoiarsk	USSR	240	∠			—

MHz	Metres	Station	Country	kW	M	M	S	N
		Kiev	Ukraine	500/240	L	_	_	_
		Mexico	Mexico	5	L	_	_	_
		Velkekostolany	Czechoslovakia	120	L	_	_	_
		Scituate	USA	50/100	_			
		Okeechobee	USA	100	_			
		Bata	Guinea	50	L	_	_	_
		Santiago	Chile	100	/	_	_	_
		Delhi	India	100	/	_	_	_
		Wien	Austria	100	_			
		Riyadh	Saudi Arabia	350		_		
		Tinang	Philippines	50				_
		Carnarvon	Australia	250				_
15.115	19.85	Jeddah	Saudi Arabia	50	L	_	_	_
		Riyadh	Saudi Arabia	350	_	_		
		Quito	Ecuador	50/100	L	_	_	_
		Lisbonne	Portugal	250	L	_	_	_
		Playa de Pals	Spain	500	_			
		Khabarovsk	USSR	100	_			
		Jigulevsk	USSR	100		_		
		Tachkent	USSR	50	_	_		
		Islamabad	Pakistan	100/250	L	_	_	_
		Nauen	Germany (E)	100				_
		Santiago	Chile	5	L	_	_	_
		Malolos	Philippines	100/250	/	_	_	_
		Poro	Philippines	50				_
		Tinang	Philippines	250	/			
		Agana	Guam	100	_			
		Ekala	Ceylon	35	/		_	_
		Okeechobee	USA	100				_
		Peking	China Rep					
15.120	19.84	London	UK	100/250	_	_		
		Merauke	Indonesia	5	L	_	_	_
		Ikorodu	Niger	100	L	_	_	_
		S M Galeria	Vatican	100	L	_	_	_
		Warszawa	Poland	15/100	L	_	_	_
		Ekala	Ceylon	35/100	L	_	_	_
		Hoerby	Sweden	350	L			
		Jerusalem	Israel	50	_		_	
		Kavalla	Greece	250	L	_	_	_
		Greenville	USA	50/250	L	_	_	_
		Delhi	India	100	_	_		
		Ismaning	Germany (W)	100				_
		Wertachtal	Germany (W)	500				
15.125	19.83	Limassol	Cyprus	20	_	_	_	
		S Gabriel	Portugal	100	L	_	_	_
		Lisbonne	Portugal	100	_	_	_	
		Salvador	Brasil	10	L	_	_	_
		Mexico	Mexico	10	L	_	_	_

MHz	Metres	Station	Country	kW	M	M	S	N
		K Wusterhausen	Germany (E)	100		—	—	
		Meyerton	South Africa	100/250	—	—	—	—
		Arganda	Spain	100	—	—	—	—
		Noblejas	Spain	350	/			
		Greenville	USA	50/500				
		Bombay	India	100	∠	—	—	—
		Taipei	China Nat					
		Hoerby	Sweden	350	—	—	—	
		Santiago	Chile	100	/	—	—	
		Ankara	Turkey	250	/			—
15.130	19.83	Simferopol	Ukraine	240/500	/			—
		Vladivostock	USSR	240	∠	—	—	
		Lampertheim	Germany (W)	100	∠		—	
		Playa de Pals	Spain	100/500		—	—	
		Scituate	USA	50/100	∠			—
		Okeechobee	USA	100			—	—
		Nauen	Germany (E)	100	∠	—		
		Wellington	New Zealand	7.5	—	—	—	
		Santiago	Chile	100	—		—	
		Kimjae	Korea (S)	250	—			
		Delhi	India	20/100	∠	—	—	—
		Hoerby	Sweden	350	∠			
		Kavalla	Greece	250	/			
		Carnarvon	Australia	250				—
15.135	19.82	Abu Zaabal	Egypt	100		—		
		Allouis	France	100/500	∠	—	—	—
		Wertachtal	Germany (W)	500	∠	—	—	—
		Julich	Germany (W)	100	∠	—	—	—
		S Paulo	Brasil	7.5	∠	—	—	
		Scituate	USA	50/100	—			
		Cyclops	Malta	250	—		—	—
		Kamalabad	Iran	250	—			
		Malolos	Philippines	50	∠	—	—	—
		Fredrikstad	Norway	120	∠	—	—	—
		Wien	Austria	100	—	—	—	
		Agana	Guam	100	/	—	—	
		Lvov	Ukraine	240	—			
		Vinnitsa	Ukraine	240	/			—
		Sofia	Bulgaria	500	—			
		Ankara	Turkey	250	/			—
15.140	19.82	London	UK	100/250	/	—	—	—
		Shepparton	Australia	10/50	—	—	—	—
		Riazan	USSR	240	∠	—	—	—
		Riga	USSR	100	—	—		
		Petropavlo Kam	USSR	240	∠			—
		Bombay	India	100	∠	—		
		Schwarzenburg	Switzerland	100/250	—	—	—	—
		Cincinnati	USA	175	—	—	—	—

MHz	Metres	Station	Country	kW	M	M	S	N
		Greenville	USA	250	/		_	_
		Santiago	Chile	100	/	_	_	_
		S Gabriel	Portugal	100	_			
		Kabul	Afghanistan					
15.145	19.81	K Wusterhausen	Germany (E)	50		_	_	
		Nauen	Germany (E)	100	∠	_	_	_
		Biblis	Germany (W)	100	_			
		Lisbonne	Portugal	100	∠	_	_	_
		Recife	Brasil	10	∠	_	_	
		Red Lion	USA	50	/	_	_	
		Agana	Guam	100	_			
		Ankara	Turkey	250	/			_
15.150	19.80	Antigua	Br W Indies	250	∠	_	_	_
		Jakarta	Indonesia	7.5	∠	_	_	_
		Santiago	Chile	10/100		_	_	_
		Julich	Germany (W)	10				_
		Wertachtal	Germany (W)	500	∠	_	_	_
		Greenville	USA	250	/	_	_	_
		Sulaibiyah	Kuwait	250		_	_	
		Islamabad	Pakistan	100	∠	_	_	_
		Sackville	Canada	50	∠	_	_	
		Minsk	Bielorussia	240/500	∠	_	_	_
		Erevan	USSR	240		_	_	
		Serpukhov	USSR	240		_		
		Omsk	USSR	240/500	∠	_	_	_
		Tinang	Philippines	250		_	_	
15.155	19.80	Allouis	France	100/500	∠	_	_	_
		Tinang	Philippines	50/250	∠	_	_	_
		S Paulo	Brasil	25	∠	_	_	
		Alma Ata	USSR	240			_	
		Duchanbe	USSR	500		_		
		Armavir	USSR	240		_		
		Meyerton	South Africa	250	∠	_	_	_
		Nauen	Germany (E)	50		_	_	
		K Wusterhausen	Germany (E)	100				
		Agana	Guam	100		_	_	_
		Greenville	USA	250		_		
		Riyadh	Saudi Arabia	350				
		Wien	Austria	100		_		
		Alger	Algier					
		Tanger	Morocco	50/100		_	_	_
		Rabat	Morocco	50/100	/			
15.160	19.79	London	UK	100/250	∠		_	
		Jaszbereny	Hungary	250	∠	_	_	_
		Szekesfehervar	Hungary	20	∠	_	_	_
		Mexico	Mexico	10	∠	_	_	_
		Greenville	USA	50/100	∠	_	_	_
		Cincinnati	USA	175/250	/	_	_	_
		Okeechobee	USA	100	_			

		Scituate	USA	100	∠			
		Lyndhurst	Australia	10	∠	_	_	_
		Athinai	Greece	100	∠	_	_	_
		Kavalla	Greece	250	∠	_	_	_
		Alger	Algeria	50	∠	_	_	_
		Mahe	Seychelles	30/100	∠		_	_
		Quito	Ecuador	100	∠		_	_
		Lopik	Netherlands	100			_	_
		Delhi	India	100	/	_	_	_
		Sines	Portugal	250	/	_	_	
		Tanger	Morocco	50/100				_
15.165	19.78	London	UK	250				_
		Limassol	Cyprus	100	_	_	_	
		Aligarh	India	250	/			
		Delhi	India	20/100	∠	_	_	_
		Forteleza	Brasil	5	∠	_	_	_
		S M Galeria	Vatican	100	_	_	_	
		Kobenhavn	Denmark	50	∠	_	_	_
		Peking	China Rep					
		Talata Volon	Malagasy Rep	300	_	_		
		Manzini	Swaziland	25	_		_	_
		Lopik	Netherlands	100	_			
		Nauen	Germany (E)	100		_		
		Kavalla	Greece	250			_	
		Tinang	Philippines	250	/			_
15.170	19.78	Lisbonne	Portugal	10/100	∠	_	_	_
		Lampertheim	Germany (W)	100		_		
		Fredrikstad	Norway	100		_		
		Papeete	Tahiti	20	∠	_	_	_
		Leipzig	Germany (E)	100	∠	_	_	_
		Nauen	Germany (E)	500		_		
		Tinang	Philippines	250		_	_	
		Greenville	USA	250				_
		Kimjae	Korea (S)	100	∠	_	_	_
		Vinnitsa	Ukraine	50/500	∠	_	_	_
		Tula	USSR	50	_	_		
		Blagovechtchen	USSR	240		_		
		Schwarzenburg	Switzerland	250	/	_	_	
15.175	19.77	Alger	Algeria					
		Abu Zaabal	Egypt	100				
		Fredrikstad	Norway	100/250	∠	_	_	_
		Kazan	USSR	240				_
		Kenga	USSR	240	_	_		
		Irkutsk	USSR	240/500	/	_	_	_
		Lvov	Ukraine	100	_			
		Red Lion	USA	50	_			
		Agana	Guam	100	∠	_	_	_
		Riyadh	Saudi Arabia	350			_	_
		Santiago	Chile	100			_	_

MHz	Metres	Station	Country	kW	M	M	S	N
15.180	19.76	London	UK	100/250	L	_	_	_
		Talata Volon	Malagasy Rep	300	_			
		Armavir	USSR	240	/			_
		Leningrad	USSR	100				
		Petropavlo Kam	USSR	50/240	/	_	_	_
		Delhi	India	100	L	_	_	_
		Carnarvon	Australia	100	/			
		Shepparton	Australia	100		_	_	_
		Bonaire Noord	Neth Antilles	300	L	_	_	_
		Allouis	France	100/500	/		_	_
		Beyrouth	Lebanon	100		_		
15.185	19.76	Antigua	Br W Indies	250	L	_	_	
		Ekala	Ceylon	35	_	_	_	_
		Jigulevsk	USSR	240		_	_	
		Serpukhov	USSR	100/240	L	_		_
		Simferopol	Ukraine	240		_		
		Ikorodu	Niger	100	L	_	_	_
		Wertachtal	Germany (W)	500	L	_	_	_
		Red Lion	USA	50	L	_	_	_
		Lopik	Netherlands	100	_	_	_	_
		Sines	Portugal	250	L	_	_	_
		Delhi	India	100	L	_	_	_
		Poro	Philippines	50	/			
		Noblejas	Spain	350	L	_	_	_
15.190	19.75	Brazzaville	Congo	50	L	_	_	_
		Aligarh	India	250	L	_	_	_
		Sackville	Canada	50/250	L	_	_	
		Belo Horizonte	Brasil	25	L	_	_	_
		Erevan	USSR	100		_	_	
		Tula	USSR	240	L			_
		Tchita	USSR	240		_		
		Cincinnati	USA	175		_	_	
		Bata	Guinea	50	L	_	_	_
		Riyadh	Saudi Arabia	350	_			
		Wavre	Belgium	100	L	_	_	_
		Allouis	France	100	L	_	_	_
		Schwarzenburg	Switzerland	150				_
15.1925	19.75	Varberg*	Sweden	100	L	_	_	_
15.195	19.74	London	UK	250	L	_	_	_
		Ascension	Ascension	250	L	_	_	_
		Tokyo Yamata	Japan	100	L	_	_	_
		Tanger	Morocco	35/100	L	_	_	_
		Warszawa	Poland	1	_	_	_	
		Allouis	France	100/500	/		_	_
		Wien	Austria	100		_		
		Achkhabad	USSR	500		_		
		Irkutsk	USSR	100	/			
		Kavalla	Greece		/		_	
		Greenville	USA	250	/		_	_

MHz	Metres	Station	Country	kW	M	M	S	N
		Okeechobee	USA	100			—	
		Peking	China Rep					
15.200	19.74	Singapore	Singapore	50	∠	—	—	—
		Kalatch	USSR	240	∠	—	—	—
		S M Galeria	Vatican	100		—	—	
		Allouis	France	100	∠	—	—	—
		Kathmandu	Nepal	100	∠	·	—	—
		Beyrouth	Lebanon	100	—			
		S Gabriel	Portugal	100	/	—	—	—
		Pori	Finland	100				
		Jerusalem	Israel	50				—
		Talata Volon	Malagasy Rep	300				—
		Tinang	Philippines	250				—
		Meyerton	South Africa	100	/			
15.205	19.73	London	UK	100/250	∠			
		Simferopol	Ukraine	240				—
		Greenville	USA	50/500	∠			—
		Delano	USA	100/250	∠			—
		Aligarh	India		/			
		Delhi	India	100/250	—	—	—	—
		Tanger	Morocco	35/100	∠	—	—	—
		Kavalla	Greece	250			—	
		Shepparton	Australia	100	—	—	—	—
		Carnarvon	Australia	250	/	—		
		Wavre	Belgium	100				—
		Poro	Philippines	50		—	—	
		Kimjae	Korea (S)	250		—	—	
		Ismaning	Germany (W)	100		—		
		Hoerby	Sweden	350		—		
		Karachi	Pakistan	50			—	
		Arganda	Spain	100	X	—	—	—
15.210	19.72	Asuncion	Paraguay	100	∠	—	—	—
		Moskva	USSR	240	/		—	
		Armavir	USSR	240/500		—		
		Alma Ata	USSR	240	—			
		Khabarovsk	USSR	100			—	
		Petropavlo Kam	USSR	240		—		
		Irkutsk	USSR	240	∠	—	—	—
		Abis	Egypt	100/250			—	
		Wavre	Belgium	100	∠	—	—	
		Allouis	France	100	∠	—	—	—
		Tinang	Philippines	250	/	—	—	—
		Pori	Finland	100			—	
		Delhi	India	20	/		—	—
		S M Galeria	Vatican	100			—	—
		Jerusalem	Israel	50				—
15.215	19.72	London	UK	100	∠	—	—	—
		Okeechobee	USA	100	∠	—	—	—
		Lisbonne	Portugal	100	∠	—	—	—

MHz	Metres	Station	Country	kW	M	M	S	N
		S Luiz	Brasil	2.5	L	_	_	_
		Malolos	Philippines	50	L	_	_	_
		Tinang	Philippines	250	L	_	_	_
		Alger	Algeria	100	L	_	_	_
		Agana	Guam	100				_
15.220	19.71	Meyerton	South Africa	/500 250	L	_	_	_
		Duchanbe	USSR	100/500			_	_
		Riga	USSR	240	L			_
		Lopik	Netherlands	100	L	_	_	_
		Szekesfehervar	Hungary	20	L			_
		Jaszbereny	Hungary	250			_	_
		Bonaire Noord	Neth Antilles	300	L	_	_	_
		Talata Volon	Malagasy Rep	300	L	_	_	
		Kamalabad	Iran	250	_			
		Riyadh	Saudi Arabia	350	_			
		Peking	China Rep					
		Allouis	France	100	/		_	_
		Arganda	Spain	100	/			_
15.225	19.70	Limassol	Cyprus	20	/			_
		Lisbonne	Portugal	100		_		
		Wertachtal	Germany (E)	500	_	_	_	
		Sfax	Tunisia	100			_	_
		Szekesfehervar	Hungary	20	_	_	_	
		Jaszbereny	Hungary	250	_	_	_	
		Cyclops	Malta	250	L	_	_	_
		Delano	USA	250	_	_	_	
		Greenville	USA	250				_
		Wavre	Belgium	100	_			
		Taipei	China Nat					
		Careysburg	Liberia	250				_
		Agana	Guam	100	/			
15.230	19.70	Habana	Cuba	100	L	_	_	_
		Voronej	USSR	240	L	_	_	
		Alma Ata	USSR	240	L	_	_	_
		Lyndhurst	Australia	10	L	_	_	_
		Melo	Uruguay	5	L	_	_	_
		Roma	Italy	100		_	_	
		Jerusalem	Israel	50/300	L	_	_	_
		Kabul	Afghanistan	100				
		Schwarzenburg	Switzerland	250	_	_		
		Wavre	Belgium	100/250			_	_
		Alger	Algeria					
		Tinang	Philippines	250				_
		Riyadh	Saudi Arabia	350	/			
		Karachi	Pakistan	10	/			
		Agana	Guam	100	/			
15.235	19.69	London	UK	100	L	_	_	_
		Greenville	USA	250/500	L	_	_	_

MHz	Metres	Station	Country	kW	M	M	S	N
		Delhi	India	100	∠	—	—	—
		Tokyo Yamata	Japan	100	∠	—	—	—
		Careysburg	Liberia	250	∠	—	—	—
		Bocaue	Philippines	50	—	—	—	—
		Malolos	Philippines	50	∠	—	—	—
		Tanger	Morocco	35	/			—
		Kavalla	Greece	250	∠		—	
		Lopik	Netherlands	100	∠	—	—	
		Agana	Guam	100			—	—
		Schwarzenburg	Switzerland	150	/			—
		Kampala	Uganda	250				—
15.240	19.69	Armavir	USSR	240	∠			
		Baku	USSR	500				—
		Novosibirsk	USSR	100	∠			
		Nikolaevskamur	USSR	240	/			—
		Kiev	Ukraine	240				—
		Belgrade	Yugoslavia	100	∠			—
		K Wusterhausen	Germany (E)	100	∠	—	—	—
		Nauen	Germany (E)	500	∠	—	—	—
		Lyndhurst	Australia	10	∠	—	—	—
		Karlsborg	Sweden	350	—			
		Hoerby	Sweden	350	∠	—	—	—
		Greenville	USA	250/500	∠	—	—	—
		Careysburg	Liberia	250	—			
		Jerusalem	Israel	20/300	—	—	—	
		Wavre	Belgium	100	—	—		
		Noblejas	Spain	350	∠	—	—	—
		Velkekostolany	Czechoslovakia	120	/			—
		Santiago	Chile					
15.245	19.68	London	UK	100	/		—	—
		Wertachtal	Germany (W)	500	∠	—	—	—
		Julich	Germany (W)	100	∠	—	—	—
		Sines	Portugal	250	∠	—	—	—
		Brasilia	Brasil	10	/			—
		Nikolaevskamur	USSR	120			—	—
		Simferopol	Ukraine	240			—	
		Kinshasa	Zaire	100	∠			
		Wien	Austria	100			—	
		Tanger	Morocco	35	/			—
		Quito	Ecuador	100			—	
15.250	19.67	London	UK	250			—	
		Limassol	Cyprus	20		—		
		Bucuresti	Roumania	120/250	∠	—	—	—
		Cincinnati	USA	250	/			—
		Dixon	USA	250	—	—	—	—
		Greenville	USA	250	∠	—	—	—
		Tinang	Philippines	250	∠	—	—	—
		Poro	Philippines	50	/		—	—
		K Wusterhausen	Germany (E)	100			—	—

MHz	Metres	Station	Country	kW	M	M	S	N
		Delhi	India	100	L	_	_	_
		Kampala	Uganda	250				_
		Careysburg	Liberia	250	_	_	_	
		Roma	Italy	60/100	_	_		
		Wavre	Belgium	250		_		
		Franceville	Gabon Rep	500		_		
		Noblejas	Spain	350	/			_
		Peking	China Rep					
15.255	19.67	K Wusterhausen	Germany (E)	100				_
		Lisbonne	Portugal	250	L	_	_	_
		Khabarovsk	USSR	120	_			
		Kenga	USSR	50/500	L			_
		Tachkent	USSR	100	_	_		
		Abu Zaabal	Egypt	100		_		
		Bonaire Zuid	Neth Antilles	50	/	_	_	_
		Bucuresti	Roumania	250	/	_	_	_
15.260	19.66	London	UK	100/250	_	_		
		Ascension	Ascension	250	L	_	_	
		Kazan	USSR	120	_	_		
		Baku	USSR	240/500	_	_	_	
		Simferopol	Ukraine	240	_			
		Tokyo	Japan	1	L	_	_	
		Momote	Japan	10	_			
		Greenville	USA	50/250	L	_	_	
		Scituate	USA	50	_			
		Kavalla	Greece	250	L	_	_	_
		Careysburg	Liberia	250	_			_
		Malolos	Philippines	50	L	_	_	_
		Tinang	Philippines	250	_			_
		Kamalabad	Iran	350	_			
		Sackville	Canada	50/250	L	_	_	_
		Shepparton	Australia	50/100	_	_	_	
15.265	19.65	London	UK	100	_	_		
		Kenga	USSR	100	/	_	_	
		Armavir	USSR	100	_	_		_
		Frunze	USSR	240/500	_			
		Gorkii	USSR	100	_			_
		Tachkent	USSR	500	/			
		S Paulo	Brasil	7.5/10	L	_	_	_
		Pori	Finland	100	L	_	_	_
		Jerusalem	Israel	100/300	_			
		Peking	China Rep					
15.270	19.65	London	UK	100/250	L	_	_	_
		Kranji	Singapore	100	/			
		Limassol	Cyprus	100				
		Peking	China Rep					
		Tanger	Morocco	100	L	_	_	_
		Delano	USA	250	_	_	_	_
		Scituate	USA	50			_	

MHz	Metres	Station	Country	kW	M	M	S	N
		Tokyo Yamata	Japan	200	∟	_	_	_
		Pori	Finland	100/250	∟	_	_	_
		Wien	Austria	100	_	_	_	_
		Kavalla	Greece	250	∟	_	_	
		Islamabad	Pakistan	10	/			
		Delhi	India	100	/			
		V of Democratic Kampuchea						
15.275	19.64	Julich	Germany (W)	100	∟	_	_	
		Wertachtal	Germany (W)	500	∟	_	_	_
		Montevideo	Uruguay	10	∟	_	_	_
		Riazan	USSR	100	∟	_	_	_
		Bonaire Zuid	Neth Antilles	50/80	∟	_	_	_
		Warszawa	Poland	100	∟	_	_	_
		Malolos	Philippines	50	/			
		Delhi	India	50				_
15.280	19.63	London	UK	100	/			
		Kranji	Singapore	100	/			_
		Greenville	USA	250/500	∟	_	_	_
		Redwood City	USA	50	∟	_	_	_
		Brazilia	Brasil					
		Armavir	USSR	240			_	_
		Krasnoiarsk	USSR	240	/		_	
		Khabarovsk	USSR			/		
		Delhi	India	50	_	_	_	_
		Malolos	Philippines	50/100	_	_	_	_
		Wavre	Belgium	250	_			
		Wellington	New Zealand	7.5			_	_
		Lopik	Netherlands	100	/			
		Allouis	France	100				
15.285	19.63	Tema	Ghana	100	∟	_	_	_
		Ejura	Ghana	250	∟	_	_	_
		K Wusterhausen	Germany (E)	100		_		_
		Tanger	Morocco	35		_		
		Szekesfehervar	Hungary	20		_	_	
		Jaszbereney	Hungary	250		_	_	
		Malolos	Philippines	50	∟	_	_	_
		Dacca	Bangladesh	100	∟	_	_	_
		Irkutsk	USSR	50				
		Meyerton	South Africa	100	/			
		Wavre	Belgium					
15.290	19.62	Shepparton	Australia	100	∟	_	_	_
		S Fernando	Argentina	10	∟	_	_	_
		Tinang	Philippines	35/250	∟	_	_	_
		Poro	Philippines	100	/			
		Noblejas	Spain	350	∟	_	_	_
		Playa de Pals	Spain	500	∟	_	_	
		Lisbonne	Portugal	50	∟			

MHz	Metres	Station	Country	kW	M	M	S	N
		Lampertheim	Germany (W)	50/100		—	—	—
		Holzkirchen	Germany (W)	10	/			
		Biblis	Germany (W)	100	/			—
		Santiago	Chile	100	—			
		Allouis	France	500		—		
15.295	19.61	Kajang	Malaysia					
		Kazan	USSR	100			—	—
		Voronej	USSR	240	—	—	—	—
		Khabarovsk	USSR	100	L	—		—
		Maputo	Mozambique	100	L	—		—
		Kajang	Malaysia	100	L	—	—	—
		Quito	Ecuador	100	L	—	—	—
		Wien	Austria	100	—			
		Kabul	Afghanistan					
15.300	19.61	Bocaue	Philippines	50	L	—	—	—
		Tokyo Yamata	Japan	100	L	—	—	—
		Quito	Ecuador	100		—	—	—
		Allouis	France	500	L	—	—	—
		Santiago	Chile	100	—			
		Novosibirsk	USSR	100			—	—
		Jerusalem	Israel	300			—	—
		Peking	China Rep					
		Tripoli	Libya					
15.305	19.60	Schwarzenburg	Switzerland	150/250	L	—	—	—
		Sottens	Switzerland	500	L	—	—	—
		Sverdlovsk	USSR	100	—			
		Tbilisi	USSR	100	L			—
		Voronej	USSR	100			—	—
		Tanger	Morocco	35/100	L		—	—
		Karlsborg	Sweden	350	—			
		Sines	Portugal	250		—		
		Red Lion	USA	50	—			
		Peking	China Rep					
		Kavalla	Greece	250		—		
		Jerusalem	Israel	300			—	
15.310	19.60	London	UK	100	/		—	—
		Kranji	Singapore	100/250				—
		Masirah	Oman	100	L	—	—	—
		Delhi	India	50	L	—	—	—
		Sofia	Bulgaria	50/145	L	—	—	—
		Conakry	Guinea	100	L	—	—	—
		Tanger	Morocco	35/100		—	—	
		Novosibirsk	USSR	100	L	—		—
		Tokyo Yamata	Japan	100/200	/	—	—	—
		Quito	Ecuador	100	L	—		—
		Malolos	Philippines	50/100		—	—	
		Bonaire Noord	Neth Antilles	300		—	—	
		Allouis	France	100		—	—	—
		Shepparton	Australia					

MHz	Metres	Station	Country	kW	M	M	S	N
15.315	19.59	London	UK	100				
		Sines	Portugal	250	∠	—	—	—
		Careysburg	Liberia	250	∠			—
		Allouis	France	100/500	∠	—	—	—
		Kamalabad	Iran	100				
		Bonaire Noord	Neth Antilles	300	∠		—	—
		Roma	Italy	100				
		Greenville	USA	50	/		—	
		Arganda	Spain	100	/		—	—
15.320	19.58	London	UK	100/250	∠	—	—	—
		Armavir	USSR	100	∠			—
		Kazan	USSR	100				
		Shepparton	Australia	50	∠		—	—
		Wertachtal	Germany (W)	500	∠	—	—	—
		Julich	Germany (W)	100	∠	—	—	—
		Bonaire Noord	North Antilles	50/300	—	—	—	—
		Careysburg	Liberia	250	/		—	—
		Leipzig	Germany (E)	100		—		
		Nauen	Germany (E)	100	/		—	—
		Wien	Austria	100	∠	—	—	—
		Santiago	Chile	100			—	
		Sines	Portugal	250				—
		Tinang	Philippines	250	/			—
15.325	19.58	Karachi	Pakistan	50	∠		—	—
		Islamabad	Pakistan	250	∠	—	—	—
		Simferopol	Ukraine	120			—	—
		Sackville	Canada	50/250	∠	—	—	—
		Mahe	Seychelles	50	∠	—	—	—
		S Paulo	Brasil	1	∠	—	—	—
		Tokyo Yamata	Japan	100		—		
		Lopik	Netherlands	100	∠	—	—	—
		Kampala	Uganda	250				
		Athinai	Greece	100	/			
15.330	19.57	Cincinnati	USA	175/250	∠	—	—	—
		Dixon	USA	100	∠	—	—	—
		Ivanofrankovsk	Ukraine	240		—		
		Starobelsk	Ukraine	240				
		Tachkent	USSR	240	∠			—
		Kursk	USSR	240				
		Sverdlovsk	USSR	100		—	—	
		Roma	Italy	100	∠	—	—	—
		Jerusalem	Israel	20/300	—	—	—	—
		Tanger	Morocco	100	/		—	
		Allouis	France	100	—		—	
		Pori	Finland	250	—			
		Kavalla	Greece	250	∠	—	—	—
		Careysburg	Liberia	250	/	—		—
		Kimjae	Korea (S)	250		—		

MHz	Metres	Station	Country	kW	M	M	S	N
		Sofia	Bulgaria	50	–			
		Riyadh	Saudi Arabia	350				–
		Arganda	Spain	100	/			–
15.335	19.56	Madras	India	100	∠	–	–	–
		Aligarh	India	250	/			
		Delhi	India	100/250	–	–	–	–
		Kimjae	Korea (S)	250	∠		–	–
		Wien	Austria	100	∠	–	–	–
		Bogota	Colombia	25	∠	–	–	–
		Abu Zaabal	Egypt	100			–	
		Bucuresti	Roumania	250	∠	–	–	–
		Dacca	Bangladesh	100			–	–
		Arganda	Spain	100	X	–	–	–
		Tanger	Morocco	100				–
		Novosibirsk	USSR	100	/			
15.340	19.56	Biblis	Germany (W)	100			–	
		Holzkirchen	Germany (W)	10	–			
		Lampertheim	Germany (W)	100	∠	–	–	
		Playa de Pals	Spain	250/500		–	–	–
		Lisbonne	Portugal	250				
		S Gabriel	Portugal	100	–			
		Habana	Cuba	50	∠	–	–	–
		K Wusterhausen	Germany (E)	100	∠			
		Bucuresti	Roumania	120	∠	–		–
		Roma	Italy	100	/	–	–	–
		Beyrouth	Lebanon	100				–
15.345	19.55	Gral Pacheo	Argentina	50	∠	–	–	–
		Sebaa Aioun	Morocco	50	∠	–	–	–
		Tinang	Philippines	250	∠	–	–	–
		Jayapura	Indonesia	5	∠	–	–	–
		Sulaibiyah	Kuwait	250	∠	–	–	–
		Fredrikstad	Norway	250	∠		–	–
		Delano	USA	250	/			–
		Dixon	USA	250	–			
		Athinai	Greece	100	∠	–	–	–
		Taipei	China Nat					
		Bucuresti	Roumania	120	∠	–	–	–
		Roma	Italy	100	∠			
		Wellington	New Zealand	7.5	/		–	–
15.350	19.54	London	UK	250	–	–	–	
		Junglinster	Luxembourg	6	∠	–	–	
		Armavir	USSR	100				
		Vologda	USSR	240	∠	–	–	–
		Komsomolskamur	USSR	240	–	–	–	
		Greenville	USA	250	–	–	–	
		Kinshasa	Zaire	100	∠		–	
		Schwarzenburg	Switzerland	100/250	∠	–	–	
		Riyadh	Saudi Arabia	350	–			
		Beyrouth	Lebanon	100				–

MHz	Metres	Station	Country	kW	M	M	S	N
		Scituate	USA	50				—
		Delhi	India	50	/			
15.355	19.54	London	UK	250	/			
		Holzkirchen	Germany (W)	10		—		
		Montevideo	Uruguay	10	∟	—	—	—
		Lisbonne	Portugal	100	∟	—	—	—
		Shepparton	Australia	50/100	∟	—	—	—
		Redwood City	USA	250	—	—	—	—
		Lyndhurst	Australia	10		—	—	—
15.360	19.53	Kranji	Singapore	250	/			—
		Rabat	Morocco	50	/			
		Tanger	Morocco	50	—	—	—	—
		Moskva	USSR	100/240	∟	—	—	—
		Irkutsk	USSR	50	—			
		Allouis	France	500	∟	—	—	—
		Kimjae	Korea (S)	250	/		—	—
		Tinang	Philippines	50/250			—	—
		Quito	Ecuador					
15.365	19.52	Las Mesas	Canary Is	50	∟	—	—	—
		Delano	USA	240	—	—	—	—
		Bucuresti	Roumania	18/120	∟	—	—	—
		Sackville	Canada	50		—	—	—
		Tinang	Philippines	50/250	—	—	—	—
		Kabul	Afghanistan	100	—	—	—	
		Noblejas	Spain	350		—		
		Riyadh	Saudi Arabia	350		—		
15.370	19.52	Limassol	Cyprus	20/100			—	
		Lisbonne	Portugal	25	/			
		Lampertheim	Germany (W)	100		—	—	
		Holzkirchen	Germany (W)	10	/			—
		Playa de Pals	Spain	250/500	/	—	—	—
		Rio de Janeiro	Brasil	10	∟	—	—	—
		Tula	USSR	100	—			
		Riyadh	Saudi Arabia	350	/			
		Delhi	India	100	/			
		Grenada	Br W Indies					
15.375	19.51	Armavir	USSR	240		—	—	
		Jigulevsk	USSR	100			—	
		Kazan	USSR	100	∟			
		Leningrad	USSR	220	—			
		Moskva	USSR	240	/			
		Kenga	USSR	100	∟			—
		Quito	Ecuador	50/100	∟	—	—	—
		Abu Zaabal	Egypt	100				
		Greenville	USA	500	/	—	—	—
		Noblejas	Spain	350	/			—
15.380	19.51	Limassol	Cyprus	20	/	—	—	
		Kranji	Singapore	100	/			
		Bucuresti	Roumania	15/250	∟	—	—	—

155

MHz	Metres	Station	Country	kW	M	M	S	N
		Biblis	Germany (W)	100	/		_	_
		Lampertheim	Germany (W)	100	_		_	_
		Lisbonne	Portugal	100	/			
		Playa de Pals	Spain	250/500	_	_		
		Wellington	New Zealand	7.5	_	_	_	
		Wien	Austria	100	/		_	_
		Quito	Ecuador					
		Meyerton	South Africa	100	/			
15.385	19.50	Bocaue	Philippines	35/50	_	_		
		Roma	Italy	60/100	L	_	_	_
		Armavir	USSR	240	L			
		Tchita	USSR	120	L	_		
		Gorkii	USSR	240		_	_	
		Novosibirsk	USSR	240		_		
		Mexico	Mexico	10/50				
		Cyclops	Malta	250	_			
		Sofia	Bulgaria	145/500	L	_	_	_
		Lopik	Netherlands	100				_
		Talata Volon	Malagasy Rep	300				_
		Wien	Austria	100	_			
		Schwarzenburg	Switzerland	150	/			_
15.390	19.49	London	UK	100/250	L	_	_	_
		Ascension	Ascension		/			_
		Bocaue	Philippines	50	L	_	_	_
		Hoerby	Sweden	350	L	_	_	_
		Caracas	Venezuela	10	L	_	_	_
		Bucuresti	Roumania	18/120	/	_	_	
		Kabul	Afghanistan	100	_	_	_	
		K Wusterhausen	Germany (E)	100		_		
		Leipzig	Germany (E)	100	/	_	_	_
		Scituate	USA	100		_		
		Noblejas	Spain	350	/			
15.395	19.49	Arganda	Spain	100	/		_	
		Noblejas	Spain	300				
		Greenville	USA	50/500	L	_	_	_
		Tinang	Philippines	250	L	_	_	_
		Armavir	USSR	100	/			
		Tachkent	USSR	240	L	_	_	_
		Velkekostolany	Czechoslovakia	120	L	_	_	_
		Ekala	Ceylon	35	L	_	_	_
15.400	19.48	Ascension	Ascension	125/250	L	_	_	_
		Limassol	Cyprus	20/100	_	_		
		Greenville	USA	250/500	L	_	_	_
		Roma	Italy	100	L	_	_	_
		Kazan	USSR	100	_			_
		Novosibirsk	USSR	50	/			
		Careysburg	Liberia	250	L		_	_
		Dacca	Bangladesh	100	L	_	_	_
		Karachi	Pakistan	50	_			

MHz	Metres	Station	Country	kW	M	M	S	N
		S M Galeria	Vatican	100	L	_	_	
		Pori	Finland	100	/			
15.405	19.47	Khabarovsk	USSR	240			_	_
		Armavir	USSR	240	L	_	_	
		Sverdlovsk	USSR	100			_	_
		Tirane	Albania					
		Jerusalem	Israel	50/300	L	_	_	_
		Cyclops	Malta	250	L	_	_	_
		Noblejas	Spain	350	L	_	_	_
		Quito	Ecuador	100	_		_	
		Peking	China Rep					
		Mahe	Seychelles	100	/			
		Wertachtal	Germany (W)	500		_		
		Ekala	Ceylon	35				_
15.410	19.47	Greenville	USA	250	L	_	_	_
		Tinang	Philippines	250	L	_	_	_
		Malolos	Philippines	50	_			
		Wien	Austria	100	L	_	_	_
		Kigali	Rwanda	250	L	_	_	_
		Moskva	USSR	50	L			_
		Dacca	Bangladesh	100	_	_		
		Shepparton	Australia	100	L	_	_	_
		Julich	Germany (W)	100				
15.415	19.46	Greenville	USA	50/250	L	_	_	_
		Ribeirao Preto	Brasil	1/7.5	L	_	_	_
		Baku	USSR	100			_	_
		Kazan	USSR	100	_	_	_	
		Nikolaevskamur	USSR	100			_	_
		Bonaire Zuid	Neth Antilles	50/250			_	_
		Jerusalem	Israel	20/300	L	_	_	_
		Shepparton	Australia	25	/	_	_	_
		Hoerby	Sweden	350		_		
		Malolos	Philippines	50				_
		Quito	Ecuador					
15.420	19.46	London	UK	250	/	_	_	_
		Ascension	Ascension	250				_
		Antigua	Br W Indies	250	_	_		
		Limassol	Cyprus	20/100	L	_	_	_
		Tokyo Yamata	Japan	20/100	L	_	_	_
		Islamabad	Pakistan	100/250	L	_	_	_
		Karachi	Pakistan	50	L	_	_	_
		Serpukhov	USSR	100	_			
		Tchita	USSR	500				_
		Khabarovsk	USSR	100	/			
		Irkutsk	USSR	100	_			
		Krasnoiarsk	USSR	100	/			
		Peking	China Rep					
		Ribeiro Preto	Brasil	1/7.5				
		Bonaire Zuid	Neth Antilles	50	/			

MHz	Metres	Station	Country	kW	M	M	S	N
15.425	19.45	London	UK	250	_			
		Petropavlo Kam	USSR	100	_	_		
		Kenga	USSR	240	_	_		
		Simferopol	Ukraine	100	_	_		
		Perth	Australia	10/50	L	_	_	_
		Wertachtal	Germany (W)	500	L	_	_	
		Julich	Germany (W)	100	L	_	_	
		Ekala	Ceylon	100	L	_	_	_
		Allouis	France	100/500	L	_	_	_
		Jerusalem	Israel	50/300	_		_	_
		Taipei	China Nat					
		Wien	Austria	100	_			
		Tinang	Philippines	250	/			_
		Greenville	USA	50				_
15.430	19.44	Limassol	Cyprus	100	/		_	_
		Schwarzenburg	Switzerland	100/250	_	_	_	_
		Sottens	Switzerland	500	_	_	_	_
		Greenville	USA	50/250	L	_	_	_
		Cincinnati	USA	175	L	_	_	_
		Delhi	India	100	_	_	_	_
		Ismaning	Germany (W)	100	_			
		Mexico	Mexico	50	L	_	_	_
		Kimjae	Korea (S)	100	L	_	_	_
		Mahe	Seychelles	100	L	_	_	
		Nauen	Germany (E)	500	_			
		Careysburg	Liberia	200	/			_
		Pori	Finland	100	/			
15.435	19.44	London	UK	100/250	L	_	_	_
		Kranji	Singapore	100	/			_
		Dar es Salaam	Tanzania	50	_	_	_	_
		Erevan	USSR	100	/			
		Frunze	USSR	100/500	L	_	_	
		Tula	USSR	120				_
		Vinnitsa	Ukraine	100	L			
		Wien	Austria	100				
		Quito	Ecuador	100	/	_	_	_
		Kamalabad	Iran	350	_			
		Allouis	France	100			_	_
		Hoerby	Sweden	350	_			
15.440	19.43	London	UK	100/250	_			
		Scituate	USA	50/100	L	_	_	_
		Okeechobee	USA	100	L	_	_	_
		Bocaue	Philippines	25/50	L	_	_	_
		Riazan	USSR	120	L	_	_	_
		Allouis	France	100	_	_	_	_
		Bonaire Zuid	Neth Antilles	50/250	L	_	_	_
		Beyrouth	Lebanon	100	_			
		Wien	Austria	100			_	
		Careysburg	Liberia	250				_

MHz	Metres	Station	Country	kW	M	M	S	N
		Franceville	Gabon Rep	500			—	
15.445	19.42	London	UK	100/250	_	_	_	
		Careysburg	Liberia	250	∠	_	_	_
		Holzkirchen	Germany (W)	10			—	
		Lampertheim	Germany (W)	100	∠	_		_
		Playa de Pals	Spain	250/500	/		_	_
		Lisbonne	Portugal	100				
		Brasilia	Brasil	10	∠	_	_	_
		Novosibirsk	USSR	240		_	_	
15.450	19.42	Serpukhov	USSR	100				
		Manila	Philippines					
		Peking	China Rep					
15.455	19.41	Moskva	USSR	240				
15.460	19.40		USSR	100				
15.465	19.40	Jerusalem	Israel					
			USSR					
		Karachi	Pakistan					
15.470	19.39		USSR	50				
15.475	19.39	Cairo	Egypt	50				
15.480	19.38		USSR	50				
		Peking	China Rep					
15.485	19.37	Jerusalem	Israel					
			USSR					
15.490	19.37		USSR*	50				
15.497	19.36		USSR*					
15.500	19.35		USSR					
			China Rep					
15.505	19.35	Sverdlovsk	USSR	50				
15.510	19.34	Peking	China Rep					
			USSR					
15.512	19.34	Jerusalem	Israel					
15.515	19.34		USSR					
		Karachi	Pakistan					
15.520	19.33	Peking	China Rep	120				
		Dacca	Bangladesh	100				
		Karachi	Pakistan					
			USSR					
15.525	19.32		USSR					
15.530	19.32		USSR					
		Jerusalem	Israel					
15.535	19.31		USSR					
15.540	19.31		USSR*					
15.545	19.30	Jerusalem*	Israel					
			USSR					
15.550	19.29	Peking	China Rep					
15.570	19.27	Seoul	Korea (S)					
15.583	19.25	Monte Carlo*	Monaco					
15.587	19.24	Peking	China					
15.589	19.24	London*	UK					

MHz	Metres	Station	Country	kW M M S N
15.590	19.24	Peking*	China Rep	50
15.595	19.24		Germany (E)*	
15.600	19.23		USSR*	
			China Rep	
		V of Democratic Kampuchea		
15.645	19.18	Kuwait	Kuwait	
15.650	19.17	Greenville*	USA	
15.660	19.16	Monte Carlo*	Monaco	
15.665	19.15	Riyadh*	Saudi Arabia	
15.670	19.14	London*	UK	
15.685	19.13	Peking	China Rep	
		Karachi	Pakistan	
15.710	19.10	Peking	China Rep	
15.730	19.07	RFE/R Liberty*		
15.735	19.07		China Rep	
15.752	19.04	Greenville*	USA	
15.770	19.02	Greenville*	USA	
15.773	19.02	Algiers	Algeria	
15.775	19.02	RFE/R Liberty*		
15.780	19.01		USSR*	
15.790	19.00	V Malayan Revln		
15.849	18.93	London*	UK	
15.850	18.93		USSR*	
15.870	18.90		USSR	
		Lyndhurst*	Australia	
15.880	18.89	Peking	China Rep	
15.910	18.86	London*	UK	
15.920	18.84	Monrovia*	Liberia	
16.030	18.71		USSR*	
16.065	18.67	RFE/R Liberty*		
16.140	18.59		USSR*	
16.190	18.53		USSR*	
16.222	18.49	Bethany*	USA	
16.240	18.47	RFE/R Liberty*		
16.247	18.46	Tangier*	Morocco	
16.250	18.46		USSR*	
16.330	18.37		USSR*	
16.430	18.26	Greenville*	USA	
16.870	17.78		USSR	
17.135	17.51		USSR	
17.387	17.25	Delhi	India	100
17.422	17.21	Hilversum*	Netherlands	
17.445	17.20	RFE/R Liberty*		
17.450	17.19	Peking	China Rep	
17.490	17.15	Peking	China Rep	
17.515	17.13	Peking	China Rep	
17.530	17.11	Peking	China Rep	
17.535	17.11	Peking	China Rep	

MHz	Metres	Station	Country	kW	M	M	S	N
17.560	17.08		USSR					
17.580	17.06		USSR*					
17.593	17.05	Baghdad	Iraq					
17.605	17.04	Peking	China Rep					
17.630	17.02	Jerusalem	Israel					
17.635	17.01	Peking	China Rep					
17.640	17.01	Santiago	Chile					
		Karachi	Pakistan					
17.642	17.00	London*	UK					
17.650	17.00	Peking	China Rep	240				
17.665	16.98	Karachi	Pakistan					
17.670	16.98	Cairo	Egypt	100				
17.680	16.97	Peking	China Rep	240				
17.685	16.96	Jerusalem	Israel					
17.690	16.96	Cairo	Egypt					
			USSR					
17.695	16.95	London	UK					
17.700	16.95	Berlin	Germany (E)					
			USSR					
		Lopik	Netherlands					
		Jerusalem	Israel					
17.705	16.94	London	UK	100/250	L	_	_	_
		Delhi	India	50/100	L	_	_	_
		Aligarh	India	250	L	_	_	_
		K Wusterhausen	Germany (E)	100	L	_	_	_
		Nauen	Germany (E)	500		_	_	
		Nikolaevskamur	USSR	100	_		_	
		Habana	Cuba	50	L	_	_	_
		S M Galeria	Vatican	100	_			
		Tanger	Morocco	20/100	L	_	_	_
		Praha	Czechoslovakia	120	L	_	_	_
		Lampertheim	Germany (W)	100		_		
		Peking	China Rep					
		V of Democratic Kampuchea						
17.710	16.94	Cincinnati	USA	250	L	_	_	_
		Moskva	USSR	240	_	_	_	_
		Tachkent	USSR	100	/			_
		Jaszbereney	Hungary	250	_	_	_	
		Szekesfehervar	Hungary	20	_	_	_	
		Jerusalem	Israel	100/300	L	_	_	_
		Wellington	New Zealand	7.5	_			
		Wien	Austria	100	_	_		
		Tanger	Morocco	35	_			
		Malolos	Philippines	100	/			
		Poro	Philippines	50			_	
		Beyrouth	Lebanon	100			_	
		Tokyo	Japan	100				
17.715	16.93	London	UK	100/250	L	_	_	_

MHz	Metres	Station	Country	kW	M	M	S	N
		Kranji	Singapore	100	/			
		Jaszbereny	Hungary	250	_	_	_	
		Szekesfehervar	Hungary	20	_	_	_	
		Wertachtal	Germany (W)	500	∟	_	_	_
		Julich	Germany (W)	100	∟	_	_	_
		Fredrikstad	Norway	100				_
		Santiago	Chile	45/100	∟	_	_	_
		Wavre	Belgium	100/250	∟	_	_	
		Schwarzenburg	Switzerland	100	_			
		Sottens	Switzerland	500	/		_	_
		Allouis	France	500				_
		Mahe	Seychelles	100	/		_	
		Careysburg	Liberia	250	/			
		Beirut	Lebanon					
17.720	16.93	Allouis	France	500	∟	_	_	_
		Red Lion	USA	50	_	_	_	_
		Dixon	USA	250	/			
		Delano	USA	250				_
		Kazan	USSR	100	∟	_	_	_
		Khabarovsk	USSR	100	_	_		
		Petropavlo Kam	USSR	100	/		_	
		Taipei	China Nat	50				
		Wien	Austria	100	_	_	_	
		Dacca	Bangladesh	100	_	_		
		Bucuresti	Roumania	120/250				_
		Noblejas	Spain	350	/			_
17.725	16.93	Kajang	Malaysia	500	_			
		Biblis	Germany (W)	100				_
		Holzkirchen	Germany (W)	10	_			_
		Lampertheim	Germany (W)	100	∟		_	
		Playa de Pals	Spain	250			_	_
		Lisbonne	Portugal	50	∟	_	_	_
		Tokyo Yamata	Japan	100	∟	_	_	_
		Abu Zaabal	Egypt	100				_
		Wien	Austria	100	_	_		
		Shepparton	Australia	100	∟	_	_	_
		Delhi	India	100	_	_	_	_
		Praha	Czechoslovakia	120	/	_	_	_
		Santiago	Chile	100				_
		Franceville	Gabon Rep	500				_
		Kuwait	Kuwait					
17.730	16.92	Scituate	USA	50/100	∟			
		Greenville	USA	250	∟	_	_	_
		Red Lion	USA	50	/			
		Okeechobee	USA	100				_
		Armavir	USSR	100/500	_	_		
		Serpukhov	USSR	240	∟	_		
		Irkutsk	USSR	240	_	_		
		Kamalabad	Iran	350	_	_		

MHz	Metres	Station	Country	kW	M	M	S	N
		Bucuresti	Roumania	50/120	∠	—	—	—
		Wavre	Belgium					
		Julich	Germany (W)	100	—	—	—	
		Wertachtal	Germany (W)	500	/			—
		Aligarh	India	250	/			
		Dacca	Bangladesh	100				
		Leipzig	Germany (E)	100				—
17.735	16.92	Peking	China Rep					—
		Lisbonne	Portugal	50/250	∠	—	—	—
		Hoerby	Sweden	30/350	—	—	—	
		Karlsborg	Sweden	350	—	—	—	
		Noblejas	Spain	350	∠	—	—	—
		Wavre	Belgium					
		Erevan	USSR	100	∠			
		Schwarzenburg	Switzerland	150	/			
		Allouis	France	500				—
			Philippines					
		Ismaning	Germany (W)					
17.740	16.91	London	UK	100/250	∠	—	—	—
		Tinang	Philippines	250	∠	—	—	—
		Poro	Philippines	100		—	—	—
		Wavre	Belgium	20/250	∠	—	—	—
		Leipzig	Germany (E)	500	∠	—		—
		K Wusterhausen	Germany (E)	100				—
		Greenville	USA	250/500	∠	—	—	—
		Lopik	Netherlands	100	—	—	—	
		Fredrikstad	Norway	250				—
		Schwarzenburg	Switzerland	150	—	—	—	
		Achkhabad	USSR	100	—	—		
		Wien	Austria	100			—	
		Sulaibiyah	Kuwait	250	/			
		Careysburg	Liberia					
17.7425	16.91	Schwarzenburg*	Switzerland	30	—	—		
17.745	16.91	Kursk	USSR	240	∠	—	—	
		Frunze	USSR	240/500	∠	—		—
		Vladivostock	USSR	50	—			
		Abis	Egypt	250		—		
		Manzini	Swaziland	25	—	—	—	
		Wien	Austria	100	—	—	—	
		Quito	Ecuador	100		—		
17.750	16.90	Holzkirchen	Germany (W)	10		—		
		Biblis	Germany (W)	100		—		
		Lampertheim	Germany (W)	100	∠	—		
		Lisbonne	Portugal	10/50	/		—	
		Playa de Pals	Spain	250	∠		—	
		Noblejas	Spain	250	∠	—	—	—
		Poro	Philippines	100	—			
		Tinang	Philippines	250	/	—		
		Delano	USA	250	/			

163

MHz	Metres	Station	Country	kW	M	M	S	N
		Dixon	USA	250	—	—	—	—
		Islamabad	Pakistan	100	∟	—	—	—
		Karachi	Pakistan	50	∟	—	—	—
		Habana	Cuba	50	∟	—	—	—
		Bucuresti	Roumania	120	∟	—	—	—
		Sackville	Canada	250	—		—	
		Sulaibiyah	Kuwait	250	—			—
17.755	16.90	K Wusterhausen	Germany (E)	100	∟	—	—	—
		Vinnitsa	Ukraine	120	/		—	
		Erevan	USSR	500	/			
		Tachkent	USSR	100/500	∟	—	—	—
		Vladivostock	USSR	100	—			
		Roma	Italy	100	∟	—	—	—
		Quito	Ecuador	100	—		—	
		Fredrikstad	Norway	100	∟			
		Riyadh	Saudi Arabia	350	∟	—	—	
		Karachi	Pakistan	50	—			
		Santiago	Chile	100		—	—	
		Tokyo Yamata	Japan	100	/	—	—	—
		Pori	Finland	250		—		
		Greenville	USA					
		Shepparton	Australia	100	/			
17.760	16.89	London	UK					—
		Biblis	Germany (W)	100	—			
		Lampertheim	Germany (W)	100	/	—	—	—
		Playa de Pals	Spain	250		—	—	—
		Lisbonne	Portugal	10		—		
		Lvov	Ukraine	240		—	—	
		Sackville	Canada	50/250	—		—	
		Riyadh	Saudi Arabia	350		—	—	
		Wien	Austria	100		—		
		Quito	Ecuador	100		—		
		Tanger	Morocco					
		Delhi	India	100	/			
17.765	16.89	Nikolaevskamur	USSR	100		—	—	
		Tula	USSR	100	∟	—	—	—
		Duchanbe	USSR	50		—	—	
		Dixon	USA	100	∟	—	—	—
		Kigali	Rwanda	250	∟	—	—	—
		Wertachtal	Germany (W)	500	∟	—	—	—
		Julich	Germany (W)	100	—			
		Poro	Philippines	100	∟	—	—	—
		Wavre	Belgium	100	∟	—		
		Alger	Algeria					
		Allouis	France	100	—	—	—	
		Beyrouth	Lebanon	100	—			
		Mexico City	Mexico					
17.770	16.88	London	UK	100	—			
		Masirah	Oman	100	/	—	—	

MHz	Metres	Station	Country	kW	M	M	S	N
		Lisbonne	Portugal	50				—
		Playa de Pals	Spain	100/250	L	—	—	
		Lampertheim	Germany (W)	100	—			
		Roma	Italy	60/100	L	—	—	—
		Wien	Austria	100	L	—	—	—
		Abu Ghraib	Iraq	100	—	—	—	—
		Hoerby	Sweden	350	—	—		
		Karlsborg	Sweden	350	—			
		Wavre	Belgium	250				
		Riyadh	Saudi Arabia	350	L	—	—	
		Julich	Germany	100		—		
		Carnarvon	Australia	250	—			
		Sottens	Switzerland	500	/			
		Schwarzenburg	Switzerland	100				—
		Wellington	New Zealand					
17.775	16.88	Vinnitsa	Ukraine	500	/			
		Armavir	USSR	100			—	
		Riazan	USSR	240	L			—
		Frunze	USSR	240/500	—			—
		Khabarovsk	USSR	240				—
		Krasnoiarsk	USSR	500				—
		Greenville	USA					
		Allouis	France	500	L	—	—	
		Jaszbereny	Hungary	250	—			
		Praha	Czechoslovakia	120	/			—
		Delhi	India	50	/			
		Malolos	Philippines	100	/			
17.7775	16.88	Varberg	Sweden	100	—			
17.780	16.87	London	UK	100	L		—	—
		Lvov	Ukraine	120	L	—	—	—
		Aligarh	India	250	L	—	—	—
		Szekesfehervar	Hungary	20	—	—	—	
		Meyerton	South Africa	500	L	—	—	
		Julich	Germany (W)	100	L	—	—	—
		Wertachtal	Germany (W)	500	L	—	—	—
		Delano	USA	200/250	L	—	—	—
		Careysburg	Liberia	250	—	—	—	
		Lampertheim	Germany (W)	100	—			
		Sackville	Canada	50/250		—		
		Poro	Philippines	50	/			—
		Tinang	Philippines	50/250	/			—
		Roma	Italy	60/100	L	—	—	—
		Kavalla	Greece	250	/	—	—	—
17.785	16.87	Abis	Egypt	250			—	
		Frunze	USSR	50	L			—
		Ivanofrankovsk	Ukraine	240				—
		Greenville	USA	250/500	L	—	—	—
		Okeechobee	USA	100			—	—
		Scituate	USA	100		—		—

165

MHz	Metres	Station	Country	kW	M	M	S	N
		Jaszbereny	Hungary	250	_	_	_	
		Szekesfehervar	Hungary	20	_	_	_	
		Bucuresti	Roumania	250	_			_
		Pori	Finland	250	_			
		Leipzig	Germany (E)	100		_		
		Santiago	Chile	100			_	
		S M Galeria	Vatican	100			_	
		Kavalla	Greece	250	/			
		Allouis	France	500	/			_
17.7875	16.87	Varberg	Sweden	100	_	_	_	_
17.790	16.86	London	UK	100/250	L	_	_	_
		Ascension	Ascension	250	_			
		Krasnoiarsk	USSR	50				
		Tinang	Philippines	250	L	_	_	_
		Bucuresti	Roumania	120/250	/	_		_
		Hoerby	Sweden	30/350	_	_	_	_
		Careysburg	Liberia	250	/	_	_	
		Allouis	France	100	_			
		Santiago	Chile	100	_	_		
		Schwarzenburg	Switzerland	150	X			_
		Vatican	Vatican					
		Quito	Ecuador	100				
17.795	16.86	London	UK	1000			_	
		Serpukhov	USSR	100	L	_	_	_
		Roma	Italy	100	L	_	_	_
		Fredrikstad	Norway	100	/	_	_	
		Tokyo Yamata	Japan	200	L	_	_	_
		Shepparton	Australia	50	L	_	_	_
		Careysburg	Liberia	250	_	_		
		Allouis	France	100	/			_
		Jerusalem	Israel	50/300	L	_	_	_
		Wertachtal	Germany (W)	500	/	_	_	
		Julich	Germany (W)	100		_	_	
17.800	16.85	London	UK	100	L	_	_	_
		Karachi	Pakistan	50	_	_	_	_
		Islamabad	Pakistan	250		_		
		Greenville	USA	50/250	L	_	_	_
		Julich	Germany (W)	100	L	_	_	_
		Wertachtal	Germany (W)	500	L	_	_	_
		K Wusterhausen	Germany (E)	100		_	_	
		Leipzig	Germany (E)	100		_		
		Fredrikstad	Norway	100	_		_	
		Tinang	Philippines	250	_	_		
		Baku	USSR	100	_			
		Allouis	France	100	L	_	_	_
		Taipei	China Nat					
		Tripoli	Libya	500				
		Kigali	Rwanda	250	L	_	_	_
		Santiago	Chile	100	L	_	_	_

MHz	Metres	Station	Country	kW	M	M	S	N
		Kathmandu	Nepal	100	∠		_	_
17.805	16.85	Tbilisi	USSR	100/500	∠	_	_	_
		Tallin	USSR	100				_
		Khabarovsk	USSR	100	/			_
		Irkutsk	USSR	240	_			
		Tula	USSR	240				
		Lisbonne	Portugal	50/250	∠	_	_	_
		Bucuresti	Roumania	250	∠	_	_	_
		Okeechobee	USA	100	/	_	_	
		Meyerton	South Africa	250				_
17.810	16.84	London	UK	250	∠	_	_	_
		Bocaue	Philippines	50	∠	_	_	_
		Lopik	Netherlands	100	∠	_	_	_
		Bonaire Noord	Neth Antilles	300	∠	_	_	_
		Talata Volon	Malagasy Rep	300	∠	_	_	_
		Warszawa	Poland	1	_	_	_	
		Tokyo	Japan					
17.815	16.84	London	UK	100				_
		Roma	Italy	60/100	∠	_	_	_
		S Paulo	Brasil	10	∠	_	_	_
		Tchita	USSR	500	/			_
		Frunze	USSR	50	_			_
		Simferopol	Ukraine	240		_	_	
		Bucuresti	Roumania	18/250	∠	_	_	
		Jerusalem	Israel	50/300	∠	_	_	_
		Wien	Austria	100	_	_		
		Talata Volon	Malagasy Rep	300	_	_	_	
		K Wusterhausen	Germany (E)	100				_
		Santiago	Chile	100	/			
		Sottens	Switzerland	500	X			
		Greenville	USA					
17.820	16.84	Kiev	Ukraine	240	_	_	_	
		Kazan	USSR	240	/			_
		Sackville	Canada	50/250	∠	_	_	_
		Dixon	USA	200	_	_	_	_
		Greenville	USA	250	/			_
		Sofia	Bulgaria	50	_	_	_	_
		Karachi	Pakistan	50	/	_	_	
		Sines	Portugal	250				_
		Tinang	Philippines	250	/	_		
		Sulaibiyah	Kuwait	250				
17.825	16.83	Tokyo Yamata	Japan	100/200	∠	_	_	_
		Baku	USSR	240/500		_	_	
		Frunze	USSR	100/500	∠	_	_	_
		Sofia	Bulgaria	50	∠	_	_	_
		Julich	Germany (W)	100	∠	_	_	_
		Wertachtal	Germany (W)	500	∠	_	_	_
		Cyclops	Malta	250	∠	_	_	_
		S M Galeria	Vatican	100	∠	_	_	_

MHz	Metres	Station	Country	kW	M	M	S	N
		Bucuresti	Roumania	120/250	∟	—	—	—
		Pori	Finland	250	/			
17.830		Ascension	Ascension	250	∟	—	—	—
		Antigua	Br W Indies	250	/		—	—
		Schwarzenburg	Switzerland	150/250	∟	—	—	—
		Islamabad	Pakistan	250	∟	—	—	—
		Athinai	Greece	100	∟	—	—	—
		Agana	Guam	100	∟	—	—	—
		Bucuresti	Roumania	120	/		—	
		Sackville	Canada	250	/			—
17.835	16.82	Lisbonne	Portugal	50	∟	—	—	—
		Vinnitsa	Ukraine	240			—	—
		Moskva	USSR	240	∟			—
		Krasnoiarsk	USSR	100	—			
		Bucuresti	Roumania	120	∟	—	—	—
		Peking	China Rep					
17.840	16.82	Antigua	Br W Indies	250		—		
		Ascension	Ascension	250		—	—	—
		Armavir	USSR		/			
		Frunze	USSR	240/500	∟			—
		Dixon	USA	200/250	/	—	—	—
		Delano	USA	200/250	∟	—	—	—
		Velkekostolany	Czechoslovakia	120	∟	—	—	—
		S M Galeria	Vatican	100	∟	—	—	—
		Bucuresti	Roumania	120	/			—
		Schwarzenburg	Switzerland	150		—	—	
		Wien	Austria	100		—	—	—
		Fredrikstad	Norway	120	∟	—	—	
		Pori	Finland	250		—	—	
		Sackville	Canada	50		—		
17.845	16.81	Scituate	USA	50/100	∟		—	
		Okeechobee	USA	100	∟	—	—	—
		Wertachtal	Germany (W)	500	—	—	—	—
		Nikolaevskamur	USSR	240		—		
		Sverdlovsk	USSR	100		—		
		Lvov	Ukraine	240		—	—	
		Lampertheim	Germany (W)	100				—
		Islamabad	Pakistan	250	∟	—	—	—
		Karachi	Pakistan	50	∟	—	—	—
		Kavalla	Greece	250	/	—		
		Allouis	France	500	/		—	—
		Jerusalem	Israel	50				—
		S M Galeria	Vatican	100	/		—	
17.850	16.81	Armavir	USSR	240	—	—		—
		Alma Ata	USSR	100		—	—	—
		Khabarovsk	USSR	100	/		—	
		Allouis	France	500	∟	—		—
		Bucuresti	Roumania	50/120	/	—	—	—
		Ekala	Ceylon	35	∟	—	—	—

168

MHz	Metres	Station	Country	kW	M	M	S	N
		Jerusalem	Israel	50				—
		Wertachtal	Germany (W)	500				
		Tinang	Philippines	250	/			—
17.855	16.80	London	UK	250	∠	—	—	—
		Talata Volon	Malagasy	300	∠	—	—	—
		Careysburg	Liberia	250	∠	—	—	—
		Greenville	USA	50/250	∠	—	—	—
		Simferopol	Ukraine	240	∠			—
		Jerusalem	Israel	50/300	∠	—	—	
		Kavalla	Greece	250	/			—
		Agana	Guam	100	∠	—	—	—
		Peking	China Rep					
		Allouis	France	100/500		—		—
		Habana	Cuba	100	/		—	
		Wien	Austria	100		—		
17.860	16.80	London	UK	100/250	/	—	—	—
		Allouis	France	100/500	∠	—	—	—
		Armavir	USSR	240		—		
		Kursk	USSR	100	∠			
		Khabarovsk	USSR	240	/			
		Tachkent	USSR	240/500		—	—	
		Vladivostock	USSR	100		—		—
		Kiev	Ukraine		∠			—
		Greenville	USA	250		—	—	
		Delhi	India	100	/			
		Aligarh	India	250		—	—	—
		Careysburg	Liberia	250	∠	—		—
		Talata Volon	Malagasy Rep	300		—	—	—
		Wellington	New Zealand	7.5	/		—	—
17.865	16.79	Scituate	USA	50/100		—		
		Warszawa	Poland	100	∠	—	—	
		Komsomolskamur	USSR	120		—		
		Delhi	India	50		—	—	—
		Allouis	France	100/500	∠	—	—	—
		Noblejas	Spain	250	/			—
		Playa de Pals	Spain	250	/			
		Biblis	Germany (W)	100				—
		Lisbonne	Portugal	10/50		—	—	—
		Quito	Ecuador	100				
		Kavalla	Greece	250	∠	—	—	—
		Poro	Philippines	100	/			—
		Wien	Austria	100	/		—	
		Careysburg	Liberia	250	/			
		Sackville	Canada	250	/			
17.870	16.79	London	UK	100/250	∠	—	—	—
		Ascension	Ascension	250	/		—	—
		Riazan	USSR	240		—		
		Khabarovsk	USSR	240	∠			—
		Tachkent	USSR	100	/			

169

MHz	Metres	Station	Country	kW	M	M	S	N
		Taichet	USSR	100				—
		Carnarvon	Australia	250	∠	—	—	—
		Shepparton	Australia	100	/	—		
		Montevideo	Uruguay	25	∠	—	—	—
		Careysburg	Liberia	250	∠	—	—	—
		Bucuresti	Roumania	120	∠	—	—	—
		Pori	Finland	250		—	—	—
		Okeechobee	USA	100				—
		Allouis	France	100				—
		Noblejas	Spain	350	/			
		Delhi	India	50	/			
17.875	16.78	Julich	Germany (W)	100	∠	—	—	—
		Wertachtal	Germany (W)	500	∠	—	—	—
		Rio de Janeiro	Brasil	7.5	∠	—	—	
		Tachkent	USSR	240/500		—	—	
		Scituate	USA	50/100	—	—	—	—
		Okeechobee	USA	100	/	—		
		Allouis	France	500	—	—		
		Jerusalem	Israel	300	/			—
		Lisbonne	Portugal	50	—			
		Pori	Finland	250		—		
		Quito	Ecuador	100				
		Fredrikstad	Norway	100	/			
		Sackville	Canada	250	/			
		Delhi	India	100				—
17.880	16.78	Ascension	Ascension	125	/			—
		Kranji	Singapore	100	/			—
		Leipzig	Germany (E)	100	—			
		Nauen	Germany (E)	500	—			
		K Wusterhausen	Germany (E)	100	∠			—
		S Gabriel	Portugal	100	∠	—	—	—
		Petropavlo Kam	USSR	240		—	—	
		Irkutsk	USSR	240	∠			—
		Tula	USSR	100		—	—	
		Kiev	Ukraine	240/500	∠			—
		Tokyo Yamata	Japan	100	∠	—	—	—
		Santiago	Chile	100	—			
		Allouis	France	100		—	—	
		Red Lion	USA	50	—			
		Peking	China Rep					
		Jerusalem	Israel	100		—		
17.885	16.77	Ascension	Ascension	125/250				
		Limassol	Cyprus	20/100	∠	—	—	
		Habana	Cuba	100	—	—	—	
		Vinnitsa	Ukraine	240	—			
		Tbilisi	USSR	100		—		
		Bogota	Colombia	25	∠	—	—	—
		Playa de Pals	Spain	250	—	—	—	
		Hoerby	Sweden	350	—			

MHz	Metres	Station	Country	kW	M	M	S	N
		Quito	Ecuador	100	/		—	
		Agana	Guam	100				—
		Lampertheim	Germany (W)	100	/			
		Dacca	Bangladesh	100	/			
		Allouis	France	500	/			
17.890	16.77	Tchita	USSR	240		—	—	
		Sverdlovsk	USSR	100	∠			—
		Lvov	Ukraine	500				—
		Taiwan	China Nat	50				
		Dacca	Bangladesh	10	∠	—	—	—	—
		Bucuresti	Roumania	120	∠	—	—	—	—
		Quito	Ecuador	100		—		—
		Shepparton	Australia	10/100	/			—
		Noblejas	Spain	350	/			—
		Agana	Guam	100	/			—
		Islamabad	Pakistan	250	/			
		Karachi	Pakistan	50	/			
17.895	16.76	Holzkirchen	Germany (W)	10		—		
		Lampertheim	Germany (W)	100	∠		—	—
		Playa de Pals	Spain	250	∠	—	—	—	—
		S Gabriel	Portugal	100	∠	—	—	—	—
		Greenville	USA	250/500			—	—
		Delano	USA	250	∠	—	—	—	—
		Lusaka	Zambia	50	∠	—	—	—	—
		S M Galeria	Vatican	100	∠	—	—	—	—
17.900	16.76	Moskva	USSR	100				
		Vatican	Vatican	100				
17.920	16.74	Cairo	Egypt	100				
18.015	16.65		USSR					
18.080	16.59	London	UK					
18.135	16.54	Dixon*	USA	50				
18.195	16.49		USSR*					
18.275	16.42	Greenville*	USA					
18.285	16.41		USSR*					
18.310	16.38		USSR*					
18.370	16.33		USSR*					
18.420	16.29		USSR					
18.460	16.25		USSR*					
18.653	16.08		USSR*					
18.782	15.97	Greenville*	USA					
18.830	15.93		USSR*					
19.130	15.68		USSR*					
19.210	15.62		USSR					
19.261	15.58	Bethany*	USA					
19.455	15.43	London	UK					
19.480	15.40	Dixon*	USA					
19.505	15.38	Greenville*	USA					
19.710	15.22	Hilversum*	Netherlands					
19.721	15.21	Greenville*	USA					

MHz	Metres	Station	Country	kW	M	M	S	N
19.725	15.21		USSR*					
19.833	15.13		USSR*					
19.845	15.12		USSR*					
19.912	15.07	Delano*	USA					
19.915	15.06	Tanger*	Morocco					
20.000	15.00	Boulder Freq Std	USA					
20.060	14.96	Greenville*	USA					
			USSR					
20.125	14.91	Greenville*	USA					
20.215	14.84	RFE*						
20.250	14.81		USSR*					
20.345	14.75	London*	UK					
20.605	14.56		USSR*					
20.710	14.49	RFE*						
21.450	13.99	Moskva	USSR					
21.455	13.98	Holskirchen	Germany (W)	10		_		
		Lampertheim	Germany (W)	20	∠	_	_	_
		Lisbonne	Portugal	50		_	_	
		Shepparton	Australia	100		_	_	
		Karachi	Pakistan	50	/			
21.460	13.98	Tula	USSR	100	/			
		Wavre	Belgium	250	∠	_	_	_
		Delano	USA	100	∠	_	_	_
		Dacca	Bangladesh	100		_	_	
21.465	13.98	Abis	Egypt	250			_	
		Nauen	Germany (E)	100/ 500	∠	_		_
		Leipzig	Germany (E)	100		_	_	
21.470	13.97	London	UK	100/250	∠	_	_	
		Greenville	USA	50		_		
		Cincinnati	USA	250	/		_	
		Wien	Austria	100		_	_	
21.475	13.97	Wavre	Belgium	100	∠	_	_	_
		Pori	Finland					
21.480	13.97	Lopik	Netherlands	100		_	_	_
		Talata Volon	Malagasy Rep	300	∠	_	_	_
		Riyadh	Saudi Arabia	350		_		
		Quito	Ecuador	100	/		_	_
21.485	13.96	Cincinnati	USA	250	∠	_	_	
		S M Galeria	Vatican	100	∠	_	_	
		K Wusterhausen	Germany (E)	100		_		
		Pori	Finland	250	/			
		Islamabad	Pakistan	250	/			
21.490	13.96	Tula	USSR	100	∠	_	_	_
		Baku	USSR	240	∠	_		_
		Bucuresti	Roumania	100/250	∠	_	_	
		Quito	Ecuador	100		_	_	
		Poro	Philippines	35	/		_	_

172

MHz	Metres	Station	Country	kW	M	M	S	N
21.495	13.96	S Gabriel	Portugal	100	∠	_	_	_
			USSR					
		Pori	Finland	250	∠	_	_	_
		Jerusalem	Israel	300	/		_	_
21.500	13.95	London	UK	250	/			
		Cincinnati	USA	175		_	_	_	_
		Greenville	USA	50/250	∠		_	_
		Wertachtal	Germany (W)	500	∠	_	_	_
		Jerusalem	Israel	300		_	_	_
		Wien	Austria	100			_	
		Riyadh	Saudi Arabia	350	/			
21.505	13.95	Duchanbe	USSR	240	∠		_	
		Erevan	USSR	240/500			_	_
		Khabarovsk	USSR	100	∠			_
		Riyadh	Saudi Arabia	350				_
		Hoerby	Sweden	350	X			_
21.510	13.95	Armavir	USSR	240			_	_
		Bogota	Colombia	25	∠	_	_	_
		Roma	Italy	100	∠	_	_	_
		Lisbonne	Portugal	50	∠	_	_	
		Poro	Philippines	50				_
21.515	13.94	Bocaue	Philippines	2	∠	_	_	_
		Simferopol	Ukraine	240/500	∠	_	_	_
		Kiev	Ukraine	240	∠			_
		Frunze	USSR	50				_
21.520	13.94	Schwarzenburg	Switzerland	150/250	∠	_	_	
		Riyadh	Saudi Arabia	350	∠		_	
		Kavalla	Greece	250	/		_	_
21.525	13.94	Jaszbereny	Hungary	250			_	_
		Szekesfehervar	Hungary	20			_	_
		Scituate	USA	50/100	∠	_	_	_
		Okeechobee	USA	100	/			_
		Shepparton	Australia	100	∠	_	_	_
		Pori	Finland	250	/			
21.530	13.93	Armavir	USSR	120	∠		_	
		Frunze	USSR	100	∠	_		
		Riyadh	Saudi Arabia	350			_	_
		Lopik	Netherlands	100			_	_
		Wien	Austria	100	/			_
21.535	13.93	Meyerton	South Africa	250	∠	_	_	_
		Greenville	USA	50			_	_
		Tokyo Yamata	Japan	100			_	_	_
		Noblejas	Spain	350	/			
21.540	13.93	Kursk	USSR	100		_		
		Nauen	Germany (E)	500			_	_
		K Wusterhausen	Germany (E)	100		_		
		Leipzig	Germany (E)	100	/	_	_	_
		Kigali	Rwanda	250	∠	_	_	_

MHz	Metres	Station	Country	kW	M	M	S	N
		Lopik	Netherlands	100		_	_	_
		Greenville	USA	250	/		_	_
21.545	13.92	Tema	Ghana	100	L	_	_	_
		Sottens	Switzerland	500	/			_
		Schwarzenburg	Switzerland	150/250	/			_
21.5475	13.92	Schwarzenburg*	Switzerland	30	/			_
21.550	13.92	London	UK	100/250	L	_	_	_
		Ascension	Ascension		/			_
		Riyadh	Saudi Arabia	350	_	_	_	_
		Sackville	Canada	50	_			
		Greenville	USA		/			
		Sines	Portugal	250	/		_	
21.5525	13.92	Varberg*	Sweden	100	/	_	_	
21.555	13.92	Roma	Italy	60/100	L		_	
		Riyadh	Saudi Arabia	350	/	_		
		Wien	Austria	100	/			
		Armavir	USSR	100	/	_	_	
		Meyerton	South Africa	250	/			
21.5575	13.92	Varberg*	Sweden	100	L	_	_	_
21.560	13.91	Roma	Italy	60/100	/	_	_	
		Julich	Germany (W)	100	L	_	_	
		Wertachtal	Germany (W)	500				
21.565	13.91	Lvov	Ukraine	100	L	_	_	
		Talata Volon	Malagasy Rep	300	/			
		Jerusalem	Israel	300	/	_	_	
		Red Lion	USA	50				
21.570	13.91	London	UK	100/250	/	_	_	
		Ascension	Ascension	250				
		Sottens	Switzerland	500	/	_	_	_
		Schwarzenburg	Switzerland	150	/		_	
		Karachi	Pakistan	50	_			
		Carnarvon	Australia	100	L	_	_	_
		Kavalla	Greece	250	_			
21.575	13.90	Moskva	USSR	100/240	L	_	_	_
		Jerusalem	Israel					
21.580	13 90	Julich	Germany (W)	100	_	_		
		Holzkirchen	Germany (W)	10				
		Allouis	France	500	L	_	_	_
		Jerusalem	Israel	300	_	_	_	
		Rabat	Morocco	50	/			
21.585	13.90	Tachkent	USSR	100	/			
		Duchanbe	USSR	100				
		Minsk	Bielorussia	240	_	_		
		Schwarzenburg	Switzerland	150	L	_	_	_
21.590	13.90	London	UK	100/250	L	_	_	_
		Ascension	Ascension	250	_	_		
		Karachi	Pakistan	50	L	_	_	
		Islamabad	Pakistan	250	L	_	_	_
		Riyadh	Saudi Arabia	350	_		_	

MHz	Metres	Station	Country	kW	M	M	S	N
		Serpukhov	USSR	240/500	/			_
		Cyclops	Malta	250	/	_	_	_
		Greenville	USA	250	/	_	_	_
		Julich	Germany (W)	100	/	_		_
		Wien	Austria	100	/			
21.595	13.89	Allouis	France	100	/			_
		Islamabad	Pakistan	100/250	/			_
		Monrovia	Liberia					
21.600	13.89	Leningrad	USSR	240	/	_	_	_
		Julich	Germany (W)	100	/	_	_	_
		Wertachtal	Germany (W)	500	/	_	_	_
		Careysburg	Liberia	50		_	_	_
		Karachi	Pakistan	50	/	_	_	
21.605	13.89	Sulaibiyah	Kuwait	250	/	_	_	_
		Poro	Philippines	50				_
21.610	13.88	London	UK	100/250		_	_	_
		Dixon	USA	50/250	/	_	_	_
		Scituate	USA	100		_		
		Greenville	USA	250	/		_	
		Tanger	Morocco	35		_		
		Hoerby	Sweden	350			_	
		Tokyo	Japan	100				
21.615	13.88	Riga	USSR	100	/	_	_	_
		Okeechobee	USA	100	/			_
		Hoerby	Sweden	350	/			_
21.620	13.88	Allouis	France	100/500	/	_	_	_
		Noblejas	Spain	350	/			
21.625	13.87	Frunze	USSR	100	/	_		_
		Jerusalem	Israel	300	/	_	_	
		Islamabad	Pakistan	250	/		_	
21.630	13.87	London	UK	250	/		_	
		Schwarzenburg	Switzerland	150	/	_	_	
		Tinang	Philippines	250	/			_
21.635	13.87	Kalinin	USSR	240	/	_	_	_
		Armavir	USSR	240			_	
21.640	13.86	London	UK	250	/	_	_	_
		Lopik	Netherlands	100	/	_	_	_
		Bonaire Noord	Neth Antilles	300	/	_	_	_
		Talata Volon	Malagasy Rep	300		_	_	
		Tokyo Yamata	Japan	100	/		_	_
21.645	13.86	Armavir	USSR	240	/	_		_
		Allouis	France	100/500	/	_	_	_
21.650	13.86	London	UK	250	/			
		Julich	Germany (W)	100	/	_	_	_
		Cyclops	Malta	250	/	_	_	_
		Karlsborg	Sweden	350		_		_
		Jerusalem	Israel	300		_	_	
		Tanger	Morocco	35			_	_
		Dixon	USA	100	/			_

MHz	Metres	Station	Country	kW	M	M	S	N
		Tripoli	Libya					
21.655	13.85	Roma	Italy	30	L	_	_	_
		Allouis	France	100	_	_		
		Fredrikstad	Norway	120	/			
		Karachi	Pakistan					
21.660	13.85	London	UK	250	/			
		Ascension	Ascension	250	/			
		Limassol	Cyprus	50/100	_	_	_	_
		Lisbonne	Portugal	50	L	_		
		Velkekostolany	Czechoslovakia	120	/	_	_	_
		Hoerby	Sweden	350	_			
		Cincinnati	USA	175	/			
21.665	13.85	Lisbonne	Portugal	50				
			USSR					
21.670	13.84	Greenville	USA	250/500	L	_	_	_
		Poro	Philippines	35	/	_	_	_
		Dacca	Bangladesh	100	/	_	_	
21.675	13.84	Allouis	France	100/500	L	_	_	
		Delhi	India	100	_	_		
		Islamabad	Pakistan	250		_	_	
21.680	13.84	Baku	USSR	100	_	_		
		Shepparton	Australia	100	/	_	_	
		Julich	Germany (W)	100	/	_	_	_
		Poro	Philippines	50	/			
21.685	13.83	Sulaibiyah	Kuwait	250	L	_	_	_
		Dacca	Bangladesh	100	L	_	_	_
		Talata Volon	Malagasy Rep	300	L	_	_	_
		Lopik	Netherlands	100	/	_	_	
		Kavalla	Greece					
21.690	13.83	Hoerby	Sweden	350	L	_	_	_
		Karlsborg	Sweden	350	/	_	_	_
		Roma	Italy	100	L	_	_	_
		Careysburg	Liberia	50		_		
		Greenville	USA	250	_			
21.695	13.83	London	UK	250	/	_	_	_
		Sackville	Canada	250	/	_	_	
		Delano	USA	100/250	/		_	
		Meyerton	South Africa	250	/			
21.700	13.82	S Gabriel	Portugal	100	L	_	_	_
		Velkekostolany	Czechoslovakia	100	L	_	_	_
		Karlsborg	Sweden	350		_	_	
21.705	13.82	Armavir	USSR	240/500	L		_	
		Allouis	France	100	/		_	
21.710	13.82	London	UK	100/250	L	_	_	_
21.715	13.82	Wien	Austria	100	/		_	_
		Bucuresti	Roumania					
21.720	13.81	Ejura	Ghana	250	L	_	_	_
		Lisbonne	Portugal	50	L	_	_	_
21.725	13.81	Tachkent	USSR	240		_	_	

MHz	Metres	Station	Country	kW	M	M	S	N
21.730	13.81	Fredrikstad	Norway	100	∟	—	—	—
		Islamabad	Pakistan	250	/	—	—	—
		Karachi	Pakistan	50	∟	—	—	—
		Allouis	France	100	∟	—	—	—
		Jerusalem	Israel	300	/			
		Rabat	Morocco					
21.735	13.80	S Gabriel	Portugal	100	∟	—	—	—
		Lisbonne	Portugal	50	/			
		Tanger	Morocco	50		—		—
		Jerusalem	Israel	50			—	
		Islamabad	Pakistan	250	/			
		Allouis	France	500	/			
21.740	13.88	Shepparton	Australia	10/50	/			—
		Delhi	India	100	/			
		Islamabad	Pakistan	100	/			
21.745	13.80	Lisbonne	Portugal	50	∟	—	—	—
		Lampertheim	Germany (W)	20		—		
		Armavir	USSR	100	∟			—
		Frunze	USSR	240/500			—	—
		Simferopol	Ukraine	240	∟			—
		Delano	USA	100		—	—	—
		Dixon	USA	50/250	/			—
		Karachi	Pakistan	50	/			
21.750	13.79	Lisbon	Portugal					
21.760	13.79	Careysburg	Liberia					
21.917	13.69	Jerusalem	Israel					
22.205	13.51		USSR					
22.770	13.18		USSR*					
22.930	13.08	London	UK					
22.970	13.06	RFE*						
23.191	12.94	London*	UK					
25.350	11.83	RFE*						
25.605	11.72	Jerusalem*	Israel	50	∟	—	—	—
25.620	11.71	Allouis	France	100	X		—	—
		Dixon	USA	100	/			
25.640	11.70	Jerusalem	Israel					
25.645	11.70	Kuwait	Kuwait					
25.650	11.70	Talata Volon	Malagasy Rep					
25.690	11.68	Lisbonne	Portugal	10	/			
25.750	11.65	Bogota	Colombia	25	∟	—	—	—
25.760	11.65	Jerusalem	Israel	50				—
25.790	11.63	Meyerton	South Africa	250	/			
25.800	11.63	Allouis	France	100	/			
25.820	11.62	Allouis	France	100				
25.880	11.59	Bethany	USA					
25.900	11.58	Allouis	France	100	/			
25.990	11.54	Delano	USA	100	/			—
26.000	11.54	Poro	Philippines	35				—
26.040	11.52	Greenville	USA	50	∟	—	—	—

MHz	Metres	Station	Country	kW	M	M	S	N
26.095	11.50	Dixon	USA	250	/		—	—
		Jerusalem	Israel	300	/			
29.705	10.10	Jerusalem	Israel					

GEOGRAPHICAL LIST OF SHORT WAVE STATIONS OF THE WORLD

	MHz
ADEN	
Aden	5.06
	5.97
	7.19
AFARS & ISSAS	
Djibuti	4.78
AFGHANISTAN	
Kabul	3.39
	4
	4.775
	6
	6.23
	7.2
	11.77
	11.805
	11.82
	11.89
	11.985
	15.14
	15.23
	15.295
	15.365
	15.39
ALBANIA	
Tirane	5.057
	5.945
	5.96
	6.18
	6.185
	6.19
	6.2
	6.21
	7.065
	7.075
	7.08
	7.09
	7.12
	7.275
	7.28
	7.29
	7.3
	9.375
	9.43
	9.48
Tirane	9.485
	9.5
	9.515
	9.75
	9.76
	9.78
	9.79
	11.845
	11.865
	11.915
	11.92
	11.935
	11.95
	11.965
	11.985
	14.27
	15.405
ALGERIA	
Alger	6.08
	6.145
	6.16
	7.055
	7.145
	7.195
	7.245
	8.063
	9.51
	9.61
	9.615
	9.64
	9.68
	9.685
	9.705
	9.76
	9.86
	11.74
	11.77
	11.8
	11.81
	11.925
	14.99
	15.155
	15.16
	15.175
	15.215
	15.23
Alger	15.773
	17.765
ANDORRA	
Andorra	6.215
ANGOLA	
Bassacongo	4.72
Huambo	4.78
	7.16
Luanda	3.345
	3.355
	3.375
	4.82
	4.985
	5.61
	5.96
	6.175
	7.235
	7.245
	7.265
	9.535
	9.66
	11.955
Saurimo	4.86
Silva Porto	4.896
Unknown	4.79
ARGENTINA	
Buenos Aires	5.882
Hurlingham	6.09
	9.76
	11.78
Gral Pacheo	6.06
	9.69
	11.71
	15.345
Lomas Mirador	5.985
	9.74
	11.88
Malargue	6.16
Mendoza	6.18
S Fernando	6.12
	9.71
	11.755
	15.29

	MHz
ASCENSION	
Ascension	6.005
	6.055
	6.14
	7.105
	7.15
	7.27
	9.58
	9.6
	9.7
	9.76
	9.765
	9.77
	11.75
	11.77
	11.79
	11.82
	11.86
	15.105
	15.195
	15.26
	15.39
	15.4
	15.42
	17.79
	17.83
	17.84
	17.87
	17.88
	17.885
	21.55
	21.57
	21.59
	21.66
AUSTRALIA	
Brisbane	4.92
	9.66
Carnarvon	6.005
	6.035
	7.24
	9.56
	9.64
	9.7
	11.705
	11.715
	11.76
	11.775
	11.8
	11.835

	MHz
Carnarvon	11.855
	11.865
	11.895
	11.935
	15.11
	15.13
	15.18
	15.205
	17.77
	17.87
	21.57
Lyndhurst	5.995
	6.045
	6.15
	7.24
	7.5
	9.54
	9.68
	11.87
	11.88
	15.16
	15.23
	15.24
	15.355
	15.87
Perth	6.14
	9.61
	15.425
Shepparton	5.955
	5.97
	6.06
	6.08
	6.12
	6.865
	7.24
	9.505
	9.55
	9.56
	9.57
	9.58
	9.6
	9.67
	9.76
	9.77
	11.705
	11.72
	11.725
	11.74
	11.76
	11.79

	MHz
Shepparton	11.81
	11.82
	11.825
	11.835
	11.855
	11.88
	11.935
	15.105
	15.14
	15.18
	15.205
	15.26
	15.29
	15.31
	15.32
	15.355
	15.41
	15.415
	17.725
	17.755
	17.795
	17.87
	17.89
	21.455
	21.525
	21.68
	21.74
Sydney	6.09
Unknown	9.645
	11.87
	12.19
	12.29
AUSTRIA	
Innsbruck	6
Wien	5.925
	5.945
	5.96
	6.015
	6.155
	6.221
	7.17
	9.585
	9.605
	9.62
	9.65
	9.66
	9.725
	9.765
	9.77

	MHz		MHz		MHz
Austria — contd		Dacca	7.29	Wavre	11.8
Wien	11.72		9.5		11.85
	11.79		9.53		11.88
	11.825		9.535		11.94
	11.87		9.54		15.19
	11.895		9.55		15.205
	12.015		9.585		15.21
	15.105		9.605		15.225
	15.11		9.615		15.23
	15.135		9.68		15.24
	15.155		9.72		15.25
	15.195		9.75		15.28
	15.245		11.65		15.285
	15.27		11.725		17.715
	15.295		11.765		17.73
	15.32		11.79		17.735
	15.335		11.89		17.74
	15.38		11.895		17.765
	15.385		11.9		17.77
	15.41		15.285		21.46
	15.425		15.335		21.475
	15.435		15.4		
	15.44		15.41	BENIN	
	17.71		15.52	Cotonou	4.87
	17.72		17.72		
	17.725		17.73	BIELORUSSIA	
	17.74		17.885	Minsk	5.975
	17.745		17.89		7.26
	17.76		21.46		9.54
	17.77		21.67		11.745
	17.815		21.685		15.15
	17.84				21.585
	17.855	BELGIUM		Orcha	6.015
	17.865	Wavre	5.965		7.105
	21.47		6.01		9.655
	21.5		6.015		9.715
	21.53		6.065		11.715
	21.555		6.08		
	21.59		6.16	BOLIVIA	
	21.715		9.61	Animas	6.015
			9.615	Catavi	6.08
BANGLADESH			9.645	Choraya	4.99
Dacca	4.88		9.655	Cochabamba	5.975
	4.89		9.68	Huanuni	5.965
	6.145		9.685	La Paz	4.845
	6.155		9.73		5.005
	6.18		9.755		6.005
	7.15		11.735		6.025
	7.245		11.785		6.035
	7.28		11.79		6.055

Location	MHz
Bolivia — contd	
La Paz	6.125
	6.155
	6.185
	6.195
	9.505
	9.555
Llallagua	5.955
Oruro	6.07
Potosi	4.965
R. Nueva America	4.795
S Cruz	6.135
	6.175
	9.605
S Cruz del Sur	4.875
Sucre	5.995
	9.715
Tarija	6.145
Yacuiba	4.805
BOTSWANA	
Gaberone	3.355
	4.845
BRASIL	
Aparacida	9.635
Aquidauana	4.795
	5.025
Aranquara	3.365
Belem	4.865
Belo Horizonte	6
	6.175
	15.19
Boa Vista	4.835
Braganca	4.825
Brasilia	6.065
	9.605
	9.665
	11.78
	15.245
	15.28
	15.445
Cuiba	4.775
	5.055
Curityba	6.045
	9.545
	11.935
Florianapolis	5.975
	9.675

Location	MHz
Fortaleza	6.105
	15.165
Goiana	4.985
	4.995
	9.755
	11.735
	11.815
Lins	3.225
Londrina	4.815
Macapa	4.915
Maceio	3.325
Manaos	4.805
	9.695
Marilia	3.235
Pernambuco	3.285
Pocas de Caldas	4.885
	4.945
	9.645
P. Alegre	5.965
	6.135
	11.875
	11.915
Prudente	5.045
Recife	6.015
	6.085
	9.565
	11.825
	11.865
	15.145
Ribeirao Preto	15.415
	15.42
Rio de Janeiro	4.875
	4.905
	5.015
	5.99
	6.035
	6.065
	6.115
	6.195
	9.515
	9.61
	9.705
	9.77
	11.805
	11.885
	11.95
	15.105
	15.37
	17.875
Salvador	6.155

Location	MHz
Salvador	9.595
	11.875
	15.125
Sao Luiz	4.785
	4.976
	15.215
Sao Paulo	5.035
	5.955
	6.055
	6.095
	6.125
	6.165
	6.185
	9.505
	9.585
	9.645
	9.685
	9.745
	11.765
	11.925
	11.965
	15.135
	15.155
	15.265
	15.325
	17.815
Taubate	4.855
Uberaba	4.965
Uberlandia	3.345
Unknown	3.265
	4.925
BRUNEI	
Berakas	4.865
	7.215
BRITISH WEST INDIES	
Antigua	5.96
	6.04
	6.055
	6.085
	6.175
	6.195
	9.51
	9.545
	9.64
	9.69
	9.735
	9.765
	11.775

	MHz		MHz		MHz
British West Indies —		Sofia	11.86	Sackville	5.96
contd	11.785		11.87		5.99
Antigua	11.795		11.92		6.045
	11.81		11.97		6.065
	11.82		15.135		6.085
	11.865		15.31		6.105
	15.15		15.33		6.12
	15.185		15.385		6.14
	15.42		17.82		6.15
	17.83		17.825		6.175
	17.84				6.195
Grenada	15.05	BURMA			9.51
	15.105	Rangoon	4.725		9.53
	15.37		5.189		9.535
Montserrat	6		5.965		9.575
	6.135		5.985		9.58
	6.145		7.185		9.59
	9.545		9.73		9.605
	9.59	Unknown	4.98		9.625
	9.64				9.635
	11.705	BURUNDI			9.655
	11.785	Bujumbari	3.3		9.715
	11.91		6.14		9.73
	11.97	Cordac	4.9		9.755
					11.705
BULGARIA		CAMBODIA			11.72
Sofia	5.915	Phnom Penh	4		11.735
	5.94		4.908		11.775
	6.07		6.095		11.825
	6.085		9.695		11.845
	6.16				11.855
	7.115	CAMEROON			11.86
	7.15	Bertoua	4.75		11.905
	7.215	Buea	3.97		11.915
	7.255		6.005		11.94
	7.27	Garoura	5.01		11.945
	7.67		7.24		11.96
	9.53	Yaounde	4.85		15.15
	9.56		4.925		15.19
	9.685		4.972		15.26
	9.7		7.205		15.325
	9.705		9.745		15.365
	9.745				17.75
	9.765	CANADA			17.76
	11.72	Calgary	6.03		17.78
	11.735	CHU time	14.67		17.82
	11.75	signal			17.83
	11.765	Halifax	6.13		17.84
	11.82	Montreal	6.005		17.865
	11.85	Ottawa	7.335		17.875

183

	MHz		MHz		MHz
Canada — contd		Ekala	11.835	Santiago	15.11
Sackville	21.55		11.87		15.115
	21.695		11.935		15.125
St Johns	6.16		11.94		15.13
Toronto	6.07		11.955		15.14
Vancouver	6.08		15.115		15.15
	6.16		15.12		15.175
			15.185		15.24
CANARY ISLANDS			15.395		15.29
Las Mesas	6.045		15.405		15.3
	6.09		15.425		15.32
	9.515		17.85		17.64
	9.525				17.715
	11.81	CHAD			17.725
	11.815	Fort Lamy	4.905		17.755
	11.88	N Djamena	6.165		17.785
	15.365		7.12		17.79
					17.8
CAPE VERDE		CHILE			17.815
ISLANDS		Concepcion	6.135		
Cape Verde	3.885	Santiago	5.955	CHINA	
Islands			6.135	(NATIONALIST)	
Barlavento	3.93		6.15	Taipei	5.98
Sao Vicente	4.72		6.19		6.405
			9.51		9.51
CENTRAL AFRICAN			9.52		9.575
REPUBLIC			9.53		9.6
Bangui	4.999		9.55		9.685
	7.22		9.57		9.765
			9.63		11.745
CEYLON			9.65		11.825
Colombo	3.383		9.675		11.86
	3.395		9.685		11.905
	4.94		9.69		11.915
Ekala	4.87		9.715		15.125
	4.9		9.75		15.225
	4.945		9.755		15.345
	6.005		9.765		15.425
	6.075		11.705		17.72
	6.13		11.715		17.8
	6.15		11.72		17.89
	6.185		11.755		
	7.105		11.76	CHINA (REPUBLIC)	
	7.11		11.765	Changsha	4.99
	7.19		11.775	Foochow	4.975
	9.72		11.78	Fukien Front	3.0
	11.745		11.8	Station	
	11.755		11.855		3.2
	11.76		11.89		3.4
	11.8		11.925		3.535

184

	MHz		MHz		MHz
China (Republic) – contd		Peking	4.96	Peking	6.86
			4.98		6.88
Fukien Front Station	3.64		5.075		6.89
	4.045		5.09		6.935
	4.33		5.125		6.955
	4.38		5.135		6.995
	5.24		5.145		7.01
	5.265		5.163		7.025
	5.9		5.22		7.03
	6.765		5.25		7.035
	7.85		5.295		7.04
Heeilunkiang	4.84		5.32		7.045
Kweiyang	3.26		5.42		7.055
Lanchow	4.865		5.84		7.06
	7.325		5.85		7.065
Nanning	4.915		5.86		7.08
Peking	3.22		5.88		7.095
	3.27		5.915		7.19
	3.29		5.935		7.315
	3.36		5.95		7.33
	3.39		5.96		7.36
	3.45		5.975		7.375
	3.5		5.99		7.38
	3.66		6.175		7.385
	3.7		6.205		7.44
	3.83		6.21		7.47
	3.92		6.225		7.48
	3.95		6.26		7.504
	3.952		6.27		7.55
	3.96		6.28		7.59
	3.985		6.29		7.62
	4.02		6.32		7.66
	4.13		6.33		7.7
	4.18		6.345		7.77
	4.2		6.41		7.775
	4.25		6.43		7.78
	4.46		6.495		7.8
	4.62		6.52		7.815
	4.63		6.54		7.82
	4.76		6.55		7.827
	4.77		6.555		7.855
	4.8		6.56		7.935
	4.815		6.585		8.007
	4.85		6.59		8.24
	4.872		6.645		8.26
	4.88		6.665		8.3
	4.885		6.75		8.32
	4.895		6.79		8.345
	4.905		6.81		8.425
	4.94		6.825		8.45

	MHz		MHz		MHz
China (Republic) —	8.49	Peking	11.33	Peking	15.195
contd	8.565		11.375		15.22
Peking	8.6		11.445		15.25
	8.66		11.455		15.265
	9.02		11.5		15.27
	9.03		11.505		15.3
	9.064		11.515		15.305
	9.08		11.53		15.405
	9.17		11.533		15.42
	9.29		11.6		15.45
	9.336		11.61		15.48
	9.34		11.63		15.51
	9.365		11.635		15.52
	9.38		11.65		15.55
	9.39		11.66		15.587
	9.4		11.665		15.59
	9.417		11.675		15.67
	9.44		11.685		15.71
	9.455		11.69		15.88
	9.46		11.695		17.45
	9.47		11.705		17.49
	9.48		11.71		17.515
	9.49		11.715		17.53
	9.515		11.72		17.535
	9.525		11.725		17.605
	9.53		11.73		17.635
	9.535		11.735		17.65
	9.54		11.945		17.68
	9.55		11.98		17.705
	9.625		12.015		17.735
	9.67		12.095		17.835
	9.82		12.06		17.855
	9.85		12.08	Urumchi	3.99
	9.86		12.11		4.11
	9.88		12.12		4.19
	9.893		12.2		4.22
	9.9		12.42		4.5
	9.92		12.45		4.97
	9.94		14.82		5.03
	9.945		15.02		5.06
	9.965		15.03		5.44
	10.245		15.04		5.925
	10.26		15.045		7.05
	10.865		15.05	Wunan	3.94
	11.000		15.06	Yunnan	4.785
	11.04		15.07	Unknown	3.925
	11.1		15.08		4.41
	11.29		15.095		4.968
	11.3		15.115		6.937
			15.165		9.895

	MHz
China (Republic) —	
contd	11.92
Unknown	15.5
	15.6
	15.735
COLOMBIA	
Aranca	4.865
	4.925
Bogota	4.755
	4.955
	4.965
	5.075
	5.96
	5.97
	6.02
	6.03
	6.065
	6.075
	6.125
	6.14
	6.16
	6.18
	9.635
	9.655
	11.79
	11.795
	11.825
	13.849
	15.335
	17.885
	21.51
	25.75
Bucramanga	4.845
Cali	6.055
	6.195
Espinal	6.095
Florencia	5.035
	6.17
Gutapuri	4.915
Ibaque	4.785
	4.875
	6.04
Medellin	4.875
	5.98
	6.105
Neiva	4.945
	6.15
Pereira	5.995
	6.01

	MHz
Popayan	6.145
Sutatenza	5.095
Tumaco	6.015
Tunja	5.985
Vallendupar	4.815
Villavicencio	4.885
	5.04
	5.955
	6.115
COMORO ISLANDS	
Dzaudazi	3.33
CONGO REPUBLIC	
Brazzaville	3.232
	4.765
	4.795
	6.115
	7.105
	7.175
	9.61
	9.715
	15.19
Point Noire	4.843
COOK ISLANDS	
Rarotonga	5.045
	9.695
	11.76
COSTA RICA	
Faro del Caribe	6.175
Pt Limon	5.955
San Jose	4.832
	5.054
	6.005
	6.04
	6.05
	6.15
	9.615
	9.645
CUBA	
Havana	6.06
	9.525
	9.55
	9.655
	9.685
	9.765
	9.77

	MHz
Havana	11.725
	11.76
	11.865
	11.93
	11.97
	15.23
	15.34
	17.705
	17.75
	17.855
	17.885
CYPRUS	
Limassol	3.99
	5.99
	6.01
	6.02
	6.05
	6.07
	6.085
	6.12
	6.125
	6.13
	6.14
	6.15
	6.155
	6.18
	6.195
	7.11
	7.14
	7.19
	7.21
	7.23
	7.25
	7.26
	7.27
	7.275
	9.53
	9.54
	9.58
	9.59
	9.6
	9.605
	9.61
	9.615
	9.62
	9.64
	9.69
	9.695
	9.705

Location	MHz	Location	MHz	Location	MHz
Cyprus — contd		Velkeko-stolany	15.24	Abu Zaabal	15.255
Limassol	9.715		15.395		15.335
	9.75		17.84		15.375
	9.77		21.66		17.725
	11.705		21.7	Cairo	6.23
	11.71				7.05
	11.72	DENMARK			7.075
	11.735	Koebenhavn	6.175		9.455
	11.74		9.71		9.475
	11.75		15.165		9.495
	11.76				9.76
	11.78	DOMINICAN			9.805
	11.82	REPUBLIC			9.85
	11.85	Pt Plata	6.19		9.995
	11.865	S Domingo	4.85		11.63
	11.955		5.01		12.005
	15.105		5.965		12.045
	15.125		6.09		12.05
	15.165		6.11		12.07
	15.225		6.13		15.475
	15.25		9.505		17.67
	15.27		9.59		17.69
	15.37		11.7		17.92
	15.38	S Pedroma-	6.025	Mokattam	9.755
	15.4	coris		Unknown	11.745
	15.42	Santiago	4.807		11.795
	15.43		6.06		
	17.885		6.075	ECUADOR	
	21.66		6.12	Caracas	4.79
R Bayrak	6.285			Esmeraldas	3.38
		EGYPT		Espejo	4.679
CZECHOSLOVAKIA		Abis	9.705	Guayaquil	4.765
Prague	5.93		9.75	Quito	4.91
	6.055		11.715		4.92
	7.245		11.785		4.923
	7.345		15.21		4.93
	9.6		17.745		5.061
	9.605		17.785		6.01
	11.855		21.465		6.05
	11.98	Abu Zaabal	6.17		6.07
	17.705		7.225		6.075
	17.725		9.52		6.095
	17.775		9.62		6.13
Velkeko-	6.015		9.675		9.56
stolany	6.055		9.77		9.585
	9.505		11.78		9.605
	9.54		11.79		9.62
	9.63		11.915		9.635
	9.74		15.135		9.65
	15.11		15.175		9.665

	MHz		MHz		MHz
Ecuador — cont		ETHIOPIA		Allouis	5.955
Quito	9.68	Gedja	4.905		5.985
	9.685		5.99		5.99
	9.705		6.015		5.995
	9.715		6.185		6.01
	9.73		7.11		6.04
	9.745		7.115		6.145
	9.76		7.165		6.175
	9.765		7.18		7.16
	11.715		9.61		7.28
	11.73		9.615		7.285
	11.745		9.705		9.505
	11.8				9.51
	11.82	FINLAND			9.525
	11.83	Pori	6.12		9.535
	11.835		9.55		9.54
	11.84		9.565		9.55
	11.865		9.575		9.56
	11.9		9.585		9.605
	11.905		9.645		9.61
	11.91		9.66		9.615
	11.915		9.755		9.62
	11.945		11.735		9.66
	11.96		11.755		9.665
	15.115		11.8		9.69
	15.16		11.825		9.695
	15.245		11.835		9.705
	15.295		11.91		9.715
	15.3		15.105		9.75
	15.31		15.2		9.755
	15.36		15.21		10.356
	15.375		15.265		11.705
	15.38		15.27		11.73
	15.405		15.33		11.735
	15.415		15.4		11.745
	15.435		15.43		11.755
	17.745		17.755		11.77
	17.755		17.785		11.78
	17.76		17.825		11.79
	17.79		17.84		11.805
	17.865		17.87		11.825
	17.875		17.875		11.84
	17.885		21.475		11.845
	17.89		21.485		11.85
	21.48		21.495		11.855
	21.49		21.525		11.86
Riobamba	3.985				11.865
S Domingo	3.39	FRANCE			11.89
Sucua	4.96	Allouis	3.965		11.91

	MHz		MHz		MHz
France -- contd		Allouis	25.62	Leipzig	21.54
Allouis	11.925		25.8	Nauen	5.955
	11.93		25.82		6.01
	11.945		25.9		6.04
	11.955	Paris	6.21		6.07
	11.965		6.745		6.08
	15.135		6.876		6.195
	15.155		11.033		7.26
	15.13		12.46		9.645
	15.19				9.665
	15.195	GABON REPUBLIC			9.73
	15.2	Franceville	3.35		9.77
	15.21		4.83		11.705
	15.22		5.955		11.805
	15.28		6.03		11.825
	15.29		7.27		11.84
	15.3		9.53		11.85
	15.31		9.585		11.89
	15.315		9.695		11.92
	15.33		11.835		11.97
	15.36		11.965		15.115
	15.425		15.25		15.13
	15.435		15.44		15.145
	15.44		17.725		15.155
	17.715	Libreville	3.3		15.165
	17.72		4.775		15.17
	17.735				15.24
	17.765	GERMANY (EAST)			15.32
	17.775	Berlin	7.3		15.43
	17.785		9.5		17.705
	17.79		11.7		17.88
	17.795		11.975		21.465
	17.8		15.1		21.54
	17.845		17.7	K Wuster-	6.01
	17.85	Leipzig	6.07	hausen	6.07
	17.855		6.195		6.08
	17.86		9.62		6.115
	17.865		9.665		7.185
	17.87		9.73		7.225
	17.875		11.72		9.505
	17.885		11.96		9.51
	21.58		15.17		9.535
	21.595		15.32		9.6
	21.62		15.39		9.665
	21.645		17.73		9.755
	21.655		17.74		11.705
	21.675		17.785		11.785
	21.705		17.8		11.795
	21.73		17.88		11.81
	21.735		21.465		11.875

Station	MHz	Station	MHz	Station	MHz
Germany (East) — contd		Berlin (RIAS)	9.555	Holzkirchen	9.52
K Wuster-	11.89		9.595		9.565
hausen	11.955		9.625		9.68
	11.97		9.66		9.695
	15.105		9.68		9.705
	15.125		9.695		9.725
	15.145		9.705		9.75
	15.155		9.725		11.825
	15.24		9.751		11.875
	15.25		11.725		11.885
	15.255		11.77		11.915
	15.285		11.78		15.29
	15.34		11.815		15.34
	15.39		11.84		15.355
	17.705		11.855		15.37
	17.74		11.875		15.445
	17.755		11.885		17.725
	17.8		11.895		17.75
	17.815		11.925		17.895
	17.88		11.935		21.455
	21.485		11.97		21.58
	21.54		15.145	Ismaning	3.98
Standard	4.525		15.291		5.955
Time Signal	15.595		15.34		5.965
			15.38		6.005
GERMANY (WEST)			17.725		6.04
Berlin (RIAS)	6.005		17.75		6.06
	3.985		17.76		6.085
	3.96		17.865		6.095
	3.99	Bremen	6.19		6.11
	5.97	Holzkirchen	3.96		6.15
	5.985		3.97		6.195
	5.955		3.99		7.15
	6.01		5.97		7.155
	6.07		5.955		7.205
	6.105		5.985		7.275
	6.115		6.01		7.29
	6.135		6.105		7.727
	6.17		6.155		9.53
	7.115		6.135		9.54
	7.165		6.17		9.62
	7.18		7.155		9.65
	7.19		7.165		9.68
	7.215		7.19		9.715
	7.22		7.2		11.78
	7.245		7.22		11.805
	7.255		7.245		11.92
	7.295		7.255		15.12
	9.505		7.295		15.205
	9.52		9.505		15.43

	MHz		MHz		MHz
Germany (West) — contd		Julich	15.41	Lampertheim	11.875
Ismaning	17.735		15.425		11.885
Julich	3.995		17.715		11.895
	5.96		17.73		11.91
	5.995		17.765		11.925
	6.04		17.77		11.935
	6.065		17.78		11.97
	6.075		17.785		15.13
	6.12		17.8		15.17
	6.13		17.825		15.29
	6.145		17.875		15.34
	6.185		21.56		15.37
	6.975		21.58		15.38
	7.105		21.59		15.445
	7.15		21.6		17.705
	7.16		21.65		17.75
	7.21		21.68		17.76
	7.235	Lampertheim	3.96		17.77
	7.275		3.985		17.725
	7.285		3.99		17.78
	7.675		5.985		17.845
	7.767		5.955		17.885
	9.545		6.105		17.895
	9.565		6.17		21.455
	9.59		7.145		21.745
	9.605		7.155	Muehlacker	6.03
	9.61		7.165	Munich	11.915
	9.615		7.18	Rohrdorf	7.265
	9.63		7.19	Wertachtal	5.96
	9.64		7.2		6.065
	9.65		7.22		6.075
	9.69		7.245		6.085
	9.7		7.255		6.1
	9.765		7.295		6.12
	9.875		9.505		6.13
	11.705		9.52		6.145
	11.765		9.54		6.185
	11.785		9.555		7.105
	11.795		9.625		7.15
	11.81		9.66		7.175
	11.85		9.68		7.21
	11.865		9.705		7.235
	11.905		9.715		7.275
	13.512		9.725		7.285
	15.105		9.75		9.51
	15.135		9.77		9.545
	15.15		11.725		9.59
	15.245		11.77		9.605
	15.275		11.855		9.61
	15.32		11.865		9.64

	MHz		MHz		MHz
Germany (West)—contd		Ejura	15.285	Kavalla	7.24
Wertachtal	9.735		21.72		7.27
	9.765	Tema	3.35		7.275
	11.72		3.366		9.53
	11.765		4.98		9.54
	11.785		6.07		9.545
	11.795		6.13		9.555
	11.82		9.545		9.585
	11.85		15.285		9.615
	11.865		21.545		9.64
	11.905				9.655
	11.935	**GREECE**			9.66
	11.945	Athinai	5.955		9.68
	15.105		6.045		9.69
	15.12		6.14		9.7
	15.135		6.185		9.735
	15.15		7.125		9.76
	15.185		7.205		9.77
	15.225		7.215		11.73
	15.245		9.515		11.74
	15.275		9.53		11.76
	15.32		9.61		11.78
	15.405		9.64		11.805
	15.425		9.655		11.81
	17.715		9.675		11.83
	17.73		9.76		11.84
	17.765		11.76		11.85
	17.78		11.845		11.855
	17.795		11.955		11.875
	17.8		15.16		11.925
	17.825		15.325		15.12
	17.845		15.345		15.13
	17.85		17.83		15.16
	17.875	Kavalla	5.955		15.165
	21.5		5.985		15.195
	21.56		6.06		15.205
	21.6		6.09		15.235
Unknown	4.882		6.095		15.26
	5.83		6.14		15.27
	10.761		6.15		15.305
	10.922		6.17		15.33
	12.29		6.19		17.78
			7.125		17.785
GHANA			7.145		17.845
Accra	3.295		7.15		17.855
	4.915		7.16		17.865
	7.295		7.17		21.52
Ejura	5.99		7.18		21.57
	11.8		7.205		21.685
	11.85		7.215	Rhodos	5.965

Greece — contd

	MHz
Rhodos	6.015
	6.07
	6.941
	7.11
	7.205
Thessaloniki	7.28
	9.655
	9.71

GREENLAND

	MHz
Godthaab	3.999
	5.96
	5.98
	9.575
	11.745

GUAM

	MHz
Agana	11.73
	11.75
	11.775
	11.78
	11.795
	11.81
	11.815
	11.84
	11.85
	11.865
	11.89
	11.9
	15.115
	15.135
	15.145
	15.155
	15.175
	15.215
	15.225
	15.23
	15.235
	17.83
	17.855
	17.885
	17.89

GUATEMALA

	MHz
Guatemala City	3.3
	6.18

GUINEA

	MHz
Bata	4.925
Bata	6.12
	7.15
	7.19
	9.555
	9.585
	11.715
	11.895
	15.11
	15.19
Bissau	5.04
Conakry	4.91
	6.155
	7.125
	9.65
	11.965
	15.07
	15.31
Malabo	6.25

GUYANA

	MHz
Georgetown	3.265
Sparendum	5.98

GUYANA (FR)

	MHz
Cayenne	3.385
	4.972
	6.17

HAITI

	MHz
Cap Haitien	11.835
Port au Prince	6.195
	9.69
	11.805

HAWAII

	MHz
Honolulu Standard Freq.	5.0
	10.0
	15.0

HONDURAS (BR)

	MHz
Belmopan	3.3

HONDURAS REP

	MHz
Comayaguela	6.11
El Progreso	4.92
La Ceiba	6.135
	6.195
S Barbara	6.075
S Pedro Sula	5.965
	5.995
S Pedros Sula	6.125
	6.185
S Rosa Copan	5.96
Tegucigalpa	4.82
	6.02
	6.035
	6.05
	6.06
	6.085
	6.095
	6.165

HUNGARY

	MHz
Budapest	8.5
	9.833
Diosd	5.955
	5.965
	5.97
	5.98
	6.025
	6.08
	6.105
	6.11
	7.2
	9.585
	9.655
	11.91
Jaszbereny	5.96
	5.98
	6.0
	6.025
	6.04
	6.06
	6.08
	6.105
	6.11
	6.125
	6.16
	6.165
	7.155
	7.215
	7.275
	9.585
	9.655
	11.91
	15.16
	15.22
	15.225
	15.285
	17.71

	MHz			MHz			MHz
Hungary — contd		Bombay		7.24	Delhi		9.535
Jaszbereny	17.715			7.26			9.565
	17.775			9.525			9.575
	17.785			9.55			9.59
	21.525			9.575			9.615
Juticalpa	6.145			11.765			9.625
Szekesfeher-	5.98			11.83			9.645
var	6.0			11.87			9.675
	6.04			15.08			9.705
	6.105			15.125			9.73
	6.11			15.14			9.755
	6.115	Calcutta		4.82			9.63
	7.155			6.01			9.912
				7.21			9.95
ICELAND				9.53			10.335
Reykjavik	12.175	Delhi		3.295			11.62
				3.365			11.715
INDIA				3.905			11.725
Aligarh	7.125			3.925			11.735
	7.225			4.86			11.74
	7.28			4.96			11.745
	9.525			5.96			11.755
	9.535			6.015			11.76
	9.565			6.02			11.765
	9.59			6.05			11.77
	9.615			6.075			11.775
	9.625			6.085			11.81
	9.675			6.105			11.825
	9.705			6.12			11.835
	11.725			6.14			11.84
	11.74			6.145			11.845
	11.745			6.15			11.85
	11.77			6.16			11.855
	11.775			6.19			11.875
	11.795			7.105			11.88
	11.81			7.11			11.885
	11.815			7.12			11.895
	15.165			7.125			11.935
	15.19			7.145			11.95
	15.205			7.165			11.945
	15.335			7.195			11.965
	17.705			7.215			15.08
	17.73			7.225			15.11
	17.78			7.235			15.12
	17.86			7.26			15.13
Bhopal	3.315			7.27			15.16
	5.99			7.28			15.165
	7.18			7.29			15.18
Bombay	4.84			7.412			15.185
	6.035			9.525			15.205

Location	MHz	Location	MHz	Location	MHz
India — contd		Madras	9.59	Jajapura	9.745
Delhi	15.21		9.715		11.865
	15.235		9.75		15.345
	15.25		15.335	Jermate	3.95
	15.27	Ranchi	6.14	Makassar	9.55
	15.275		7.17		11.75
	15.28	Sibolga	4.775	Manokwari	6.185
	15.31	Simla	3.223	Medan	3.421
	15.335		6.02		4.764
	15.35	Srinagar	6.11		7.24
	15.37	Unknown	3.277	Menado	5.99
	15.43		4.85		7.295
	17.387		17.855	Merauke	7.185
	17.705				15.12
	17.725	INDONESIA		Padang	3.905
	17.76	Amboina	7.14		3.955
	17.775	Bandjarmasin	5.97		3.96
	17.86	Biak	7.21		6.19
	17.865	Bukittinggi	4.885		6.2
	17.87	Denpassar	3.945		9.51
	17.875		7.12	Pakanbaru	5.955
	21.74	Djambi	4.927	Palembang	4.855
	21.675	Jakarta	3.925	Pontianak	3.965
Gauhati	3.235		4.07		3.995
	3.375		4.72		4.005
	4.775		4.755	Semarinda	6.135
	5.97		4.77	Semarang	3.935
	6.13		4.805	Sorong	9.6
	7.15		4.875		9.66
	7.28		4.92	Surabaya	3.98
Hyderabad	4.8		4.93		4.7
	6.12		4.932		6.12
	7.14		4.96	Surakarta	3.91
	9.72		5.05	Yogyakarta	7.105
Jammu	5.96		5.88	Unknown	3.94
Kohima	6.065		5.885		
	7.105		5.98	IRAN	
Kurseong	3.355		6.045	Kamalabad	7.205
	4.895		7.11		7.215
	6.1		7.22		7.23
	7.23		7.27		7.27
Lucknow	3.205		9.68		9.535
	6.17		9.77		9.575
	7.25		11.715		9.72
Madras	4.92		11.77		9.765
	6.085		11.79		9.77
	7.16		15.15		11.735
	7.26	Jajapura	6.07		11.745
	9.51		7.165		11.77
	9.515		7.19		11.78

196

	MHz		MHz		MHz
Iran — contd		Jerusalem	9.445	Jerusalem	17.685
Kamalabad	11.865		9.495		17.7
	11.93		9.54		17.71
	15.135		9.63		17.795
	15.22		9.65		17.815
	15.26		9.74		17.845
	15.315		9.815		17.85
	15.435		9.82		17.855
	17.73		9.833		17.875
Teheran	9.022		11.602		21.495
	15.084		11.62		21.5
			11.625		21.565
IRAQ			11.635		21.575
Abu Ghraib	7.18		11.64		21.58
	11.725		11.655		21.625
	11.825		11.7		21.65
	17.77		11.705		21.73
Babel	7.18		11.715		21.735
	9.635		11.735		21.917
	9.745		11.76		25.605
	11.905		11.775		25.64
	11.925		11.78		25.76
	11.935		11.81		26.095
Baghdad	3.195		11.84		29.705
	3.242		11.845		
	3.95		11.875	ITALY	
	11.795		11.955	Caltanissetta	6.06
	17.593		11.96		7.175
Salman Pack	6.095		12.077		9.515
	6.155		15.07	Milan	6.208
	7.22		15.1	Roma	3.995
	7.24		15.105		5.0
	9.555		15.12		5.99
	9.745		15.2		6.01
	11.78		15.21		6.05
	11.785		15.23		6.06
			15.24		6.075
ISRAEL			15.265		7.235
Jerusalem	5.882		15.3		7.275
	5.9		15.305		7.29
	5.915		15.33		9.575
	7.125		15.405		9.58
	7.2		15.415		9.63
	7.395		15.425		9.71
	7.412		15.465		11.8
	7.465		15.485		11.81
	9.009		15.512		11.875
	9.355		15.53		11.905
	9.425		15.545		11.955
	9.435		17.63		15.23

	MHz
Italy — contd	
Roma	15.25
	15.315
	15.33
	15.34
	15.345
	15.385
	15.4
	17.755
	17.77
	17.78
	17.795
	17.815
	21.51
	21.555
	21.56
	21.655
	21.69
Turin	5.0

IVORY COAST

	MHz
Abidjan	4.94
	6.015
	7.215
	11.92

JAPAN

	MHz
Hiroshima	6.175
Hokkaido	3.945
Kumamoto	6.13
Matsuyama	6.005
Momote	15.26
Nagoya	9.535
Osaka	6.19
Tokyo	3.91
	3.925
	3.93
	6.155
	11.75
	15.0
	15.26
	17.71
	17.81
	21.61
Tokyo Kawagu	9.55
Tokyo Nagara	6.055
	6.115
	9.595
	9.76
Tokyo Yamata	5.99

	MHz
Tokyo Yamata	6.03
	7.14
	7.195
	7.205
	7.25
	9.505
	9.525
	9.53
	9.585
	9.605
	9.67
	9.7
	9.705
	11.705
	11.725
	11.78
	11.815
	11.84
	11.855
	11.875
	11.94
	11.95
	11.96
	15.105
	15.195
	15.235
	15.27
	15.3
	15.31
	15.325
	15.42
	17.725
	17.755
	17.825
	21.535
	21.64

JORDAN

	MHz
Amman	7.155
	9.53
	9.56
	11.92

KENYA

	MHz
Langata	4.915
Mombasa	4.885
Nairobi	4.804
	4.934
	4.95
	7.125

	MHz
Nairobi	7.14
	7.15
	7.21
	7.24
	7.295
	9.665

KOREA (NORTH)

	MHz
Hamhung	2.775
Kanggye	4.03
Pyongyang	2.765
	2.85
	3.015
	3.32
	3.56
	3.695
	3.89
	4.77
	5.88
	6.25
	6.29
	6.338
	6.4
	6.576
	6.6
	6.77
	9.42
	9.977
	11.35
	11.535
	11.568
	11.885
Sariwon	2.670
Sinuiju	2.745
Wonsan	3.03

KOREA (SOUTH)

	MHz
Kimjae	6.03
	6.17
	7.115
	7.24
	7.425
	7.25
	7.29
	9.51
	9.525
	9.535
	9.54
	9.555
	9.565

	MHz		MHz		MHz
Korea (South) —		Sulaibiyah	11.94	LESOTHO	
contd	9.575		11.98	Maseru	4.8
Kimjae	9.595		15.15		
	9.61		15.345	LIBERIA	
	9.65		17.74	Careysburg	6.035
	9.66		17.75		6.045
	9.675		17.82		6.18
	9.71		21.605		7.175
	9.72		21.685		7.195
	9.765		25.645		7.28
	11.74	Unknown	12.085		9.54
	11.78				9.615
	11.79	LAOS			9.63
	11.84	Houa Phan	4.65		9.67
	11.845		4.665		9.7
	11.85		6.2		9.715
	11.86	Laos	7.095		9.74
	11.91	Luang	4.7		11.705
	11.945	Prabang	6.975		11.71
	11.965		8.395		11.715
	11.97	Pakse	6.6		11.74
	15.13	Savannakhet	7.385		11.805
	15.17	Udomsai	6.91		11.84
	15.205	Vientiane	3.9		11.915
	15.33		4.32		15.225
	15.335		4.645		15.235
	15.36		5.16		15.24
	15.43		6.13		15.25
Kyong San	5.975		6.21		15.26
Seoul	6.24		7.48		15.315
	7.55	Xieng	4.603		15.32
	9.87	Khouang	6.675		15.33
	15.57	Unknown	8.66		15.4
Suwon	6.135				15.43
	7.15	LEBANON			15.44
	7.275	Beyrouth	5.98		15.445
	9.58		9.545		17.715
	9.64		11.785		17.74
	9.665		11.825		17.78
Unknown	3.93		11.925		17.79
			11.965		17.795
KUWAIT			15.18		17.855
Kuwait	9.84		15.2		17.86
	15.645		15.34		17.865
	17.725		15.35		17.87
Magwa	9.65		15.35		21.6
Sulaibiyah	6.055		15.44		21.69
	7.12		17.71		21.76
	9.505		17.715	Monrovia	3.227
	9.575		17.765		3.255

Location	MHz	Location	MHz	Location	MHz
Liberia — contd		Talata Volon	15.18	Penang	9.71
Monrovia	3.99		15.2	Sabah	4.975
	4.77		15.22	Sibu	6.05
	6.075		15.385	Stapok	7.145
	7.442		17.81		7.16
	9.55		17.815		7.27
	10.88		17.855		9.535
	11.83		17.86		9.605
	11.835		21.48	Tebrau	3.915
	11.86		21.565		6.195
	11.87		21.64		
	11.88		21.685	MALDIVE ISLANDS	
	11.93		25.65	Male	6.15
	11.945	Tananarive	3.232		7.225
	12.01		3.288		9.55
	12.02		3.37		
	15.09		4.985	MALI	
	15.92		5.01	Bamako	4.783
	21.595		6.135		4.825
			6.17		4.875
LIBYA			7.105		5.995
Tripoli	6.1		7.155		7.11
	6.185		7.23		7.285
	6.2				9.635
	7.165	MALAWI			11.96
	7.2	Blantyre	3.38		
	9.5	Limbe	5.995	MALTA	
	9.565			Cyclops	5.96
	9.65	MALAYSIA			5.98
	9.655	Kajang	5.965		5.99
	9.66		6.025		6.0
	11.7		6.1		6.025
	11.755		6.175		6.1
	11.795		7.295		7.12
	15.1		9.515		7.16
	15.3		9.665		7.265
	17.8		9.75		7.275
	21.65		11.9		9.53
			15.295		9.565
LUXEMBOURG			17.725		9.605
Junglinster	6.09	Koya			9.61
	15.35	Kinabalu	4.972		9.65
		Kuala Lumpur	4.845		9.67
MALAGASY REP		Kuching	4.835		9.735
Talata Volon	6.02		4.895		9.77
	7.285		4.955		11.765
	11.73		5.03		11.785
	11.735	Mira	6.06		11.795
	11.74	Penang	4.985		11.81
	15.165		7.2		11.92

Malta — contd

Station	MHz
Cyclops	15.105
	15.135
	15.225
	15.385
	15.405
	17.825
	21.59
	21.65

MARTINIQUE

Station	MHz
Fort de France	3.315
	5.995

MAURETANIA

Station	MHz
Nouakchott	4.845
	7.245
	9.61

MAURITIUS

Station	MHz
Malherbes	9.71
Mauritius	9.71
P Louis	4.85

MEXICO

Station	MHz
Cd Mante	6.09
Chihuahua	6.14
Hermosillo	6.115
Linares	5.98
Merida	6.105
Mexico City	5.985
	6.01
	6.165
	6.185
	9.515
	9.555
	9.6
	9.705
	11.74
	11.77
	11.88
	15.11
	15.125
	15.16
	15.385
	15.43
	17.765
S Luis Potosi	6.045
Sisoguichi	5.96
Tapachula	6.12
Texmelucan	6.065
Tlaxiaco	6.145
Vera Cruz	6.02

MONACO

Station	MHz
Monte Carlo	5.945
	5.965
	6.035
	6.215
	7.105
	7.125
	7.145
	7.215
	7.23
	7.245
	7.26
	7.275
	9.51
	9.525
	9.545
	9.575
	9.59
	9.61
	9.625
	9.63
	9.64
	9.65
	9.655
	9.66
	9.675
	9.725
	9.73
	9.735
	11.705
	11.71
	11.735
	11.74
	11.75
	11.78
	11.79
	11.8
	11.835
	11.855
	11.935
	11.955
	12.295
	15.583
	15.66

MONGOLIAN REPUBLIC

Station	MHz
Hailar	3.9
	6.08
Huhetot	3.93
	3.97
	4.068
	4.895
	6.84
Ikechao/ Tungsheng	4.525
Silinhot	4.952
	7.2
Tungsheng	6.045
Ulan Bator	4.08
	4.763
	4.79
	4.83
	4.85
	4.995
	5.055
	6.385
	7.26
	11.74
	11.855
	12.07

MOROCCO

Station	MHz
Rabat	6.1
	7.365
	9.615
	15.155
	15.36
	21.58
Sebaa Aioun	6.19
	7.225
	15.345
Tangier	5.46
	5.965
	6.015
	6.06
	6.08
	6.095
	6.1
	6.11
	6.13
	6.17
	6.18
	7.17
	7.18

	MHz		MHz		MHz
Morocco — contd		Tangier	19.915	Lopik	7.21
Tangier	7.19		21.61		7.24
	7.21		21.65		9.63
	7.22		21.735		9.66
	7.23				9.715
	7.24	MOZAMBIQUE			9.77
	7.265	Beira	3.37		11.72
	7.275		4.895		11.725
	7.295		6.025		11.73
	9.51		6.09		11.74
	9.53		9.635		11.845
	9.54	L Marques	3.265		11.93
	9.58		4.855		11.935
	9.615	Maputo	3.21		11.95
	9.63		3.338		11.955
	9.65		4.865		15.16
	9.66		4.925		15.165
	9.68		6.05		15.185
	9.715		6.115		15.22
	9.735		7.11		15.235
	9.76		7.24		15.28
	9.77		9.62		15.325
	10.972		11.82		15.385
	11.71		15.295		17.7
	11.715	Nampula	3.343		17.74
	11.735		7.14		17.81
	11.76		7.255		21.48
	11.77	Pemba	7.15		21.53
	11.805	Quelimane	7.145		21.54
	11.915				21.64
	11.955	NEPAL			
	11.96	Kathmandu	6.1	NETHERLANDS	
	11.965		7.105	ANTILLES	
	15.155		7.165	Bonaire Noord	6.02
	15.16		9.59		6.165
	15.195		10.010		9.59
	15.205		11.97		9.63
	15.235		15.2		9.715
	15.245		17.8		9.77
	15.27	Khumaltar	3.425		11.79
	15.285	Lalitpur	5.005		15.18
	15.305				15.22
	15.31	NETHERLANDS			15.31
	15.33	Hilversum	9.895		15.315
	15.335		17.422		15.32
	15.36		19.71		17.81
	16.247	Lopik	5.955		21.64
	17.705		6.02	Bonaire Zuid	6.19
	17.71		6.045		9.535
	17.76		6,085		9.55

	MHz		MHz		MHz
Netherlands — cont		Bonanza	9.52	Fredrikstad	11.87
Antilles		Managua	5.95		11.895
Bonaire Zuid	9.57		6.01		11.935
	9.61		11.875		15.135
	9.62	Nueva Segovia	6.1		15.17
	9.755	Ocotal	6.1		15.175
	11.705				15.345
	11.71	NIGER			17.715
	11.79	Niamey	3.26		17.74
	11.8		5.02		17.755
	11.815				17.795
	11.83	NIGERIA			17.8
	11.9	Benin	4.932		17.84
	11.925		5.965		17.875
	11.95	Calabar	6.145		21.655
	15.255	Enugu	6.025		21.73
	15.275		7.235		
	15.415	Ibadan	3.205	OMAN	
	15.42		6.05	Masirah	6.03
	15.44		7.285		6.14
		Ikorodu	7.255		6.195
NEW CALEDONIA			7.275		7.14
Noumea	3.355		11.77		7.23
	7.17		11.9		7.25
	9.505		11.925		7.275
	11.71		15.12		9.59
			15.185		11.74
NEW HEBRIDES		Jaji	6.09		11.78
Port Vila	7.26		9.57		11.91
		Jos	5.965		11.955
NEW ZEALAND		Kaduna	3.396		15.31
Wellington	6.105		6.175		17.77
	9.62	Lagos	4.99	Seeb	6.175
	9.77	Maidguri	6.1		11.89
	11.705	Sogunle	7.255		
	11.8	Sokoto	6.195	PAKISTAN	
	11.84			Islamabad	3.24
	11.85	NORWAY			3.915
	11.945	Fredrikstad	6.005		4.02
	11.96		6.015		5.01
	15.28		6.085		5.061
	15.345		6.18		6.01
	15.38		9.55		6.03
	17.71		9.59		6.18
	17.77		9.605		7.165
	17.86		9.61		7.195
			9.645		7.265
NICARAGUA			11.735		7.29
Bluefields	5.955		11.85		9.505
	6.12		11.86		9.545

	MHz		MHz		MHz
Pakistan — contd		Karachi	11.672	Panama	9.685
Islamabad	9.59		11.68		
	9.645		11.705	PAPUA NEW GUINEA	
	9.66		11.75	Alotau	6.04
	9.69		11.885	Daru	3.305
	11.705		11.915		6.08
	11.75		15.205	Milne Bay	3.36
	11.885		15.23	Port Moresby	3.925
	11.91		15.325		4.89
	11.915		15.4		9.52
	11.955		15.42		9.575
	11.97		15.465	Rabaul	3.385
	15.115		15.515		5.985
	15.15		15.52	Wewak	3.335
	15.27		15.685		6.14
	15.325		17.64		
	15.42		17.665	PARAGUAY	
	17.75		17.75	Ascuncion	6.015
	17.8		17.755		6.025
	17.83		17.8		6.11
	17.845		17.82		9.735
	17.89		17.845		11.85
	21.485		17.89		15.21
	21.59		21.455	Concepcion	11.915
	21.595		21.57	Encarnacion	11.94
	21.625		21.59		11.945
	21.675		21.6	PJ Caballero	5.995
	21.73		21.655	Villarrica	5.975
	21.735		21.73		
	21.74		21.745	PERU	
Karachi	3.86	Peshawar	3.155	Andahuaylas	5.025
	3.89		3.33	Andina	4.995
	3.965		6.08	Iquitos	4.79
	4.72	Quetta	3.273	Lima	3.24
	4.735		5.98		4.975
	6.115		7.105	Tarapoto	4.935
	6.235		7.15		
	7.085		9.735	PHILIPPINES	
	7.165	Rawalpindi	3.215	Bocaue	6.03
	7.195		4.06		6.12
	7.21		4.935		7.225
	7.265		6.13		9.505
	7.375		9.77		9.715
	9.46	Unknown	3.396		9.755
	9.465		3.925		9.765
	9.645		5.112		11.765
	9.66				11.82
	9.69	PANAMA			11.855
	9.79	David	6.045		11.89
	11.64	Panama	5.995		11.92

Location	MHz	Location	MHz	Location	MHz
Philippines — contd		Marulas	6.17	Tinang	9.63
Bocaue	15.235		9.615		9.645
	15.3	Poro	2.6		9.65
	15.385		5.995		9.67
	15.39		6.185		9.725
	15.44		7.155		9.76
	17.81		7.165		11.715
	21.515		7.19		11.73
Iba	9.715		7.235		11.75
	9.755		7.24		11.755
	11.92		7.255		11.76
Malolos	5.965		7.275		11.78
	9.515		7.285		11.805
	9.535		9.53		11.81
	9.54		9.65		11.815
	9.57		9.77		11.83
	9.58		11.715		11.925
	9.59		11.73		11.93
	9.605		11.835		11.965
	9.64		11.84		15.11
	9.69		11.93		15.115
	9.7		11.965		15.15
	9.74		15.115		15.155
	11.725		15.185		15.165
	11.8		15.205		15.17
	11.805		15.25		15.2
	11.825		15.29		15.21
	11.83		17.71		15.215
	11.87		17.74		15.23
	11.875		17.75		15.25
	11.95		17.765		15.26
	11.955		17.78		15.29
	11.97		17.865		15.32
	15.115		21.49		15.345
	15.135		21.51		15.36
	15.215		21.67		15.365
	15.235		21.68		15.395
	15.26	Tinang	5.955		15.41
	15.275		6.01		15.425
	15.28		6.11		17.74
	15.285		7.16		17.75
	15.31		7.275		17.78
	15.41		9.545		17.79
	15.415		9.555		17.8
	17.71		9.56		17.82
	17.775		9.585		17.85
Manila	3.285		9.605		21.63
	3.345		9.615	Unknown	9.575
	15.45		9.625		11.845

205

	MHz		MHz		MHz
Philippines — contd		Lisbon	9.505	Lisbon –	25.69
Unknown	15.195		9.555	S Gabriel	6.025
			9.565		6.155
POLAND			9.595		9.615
Warsaw	3.955		9.695		9.62
	5.96		9.705		9.635
	5.995		9.725		9.74
	6.035		9.75		11.8
	6.095		10.905		11.84
	6.135		11.725		11.875
	6.155		11.77		11.925
	6.195		11.78		11.935
	7.11		11.815		15.125
	7.125		11.825		15.14
	7.145		11.855		15.2
	7.18		11.875		15.34
	7.205		11.885		17.88
	7.27		11.895		17.895
	7.285		11.925		21.495
	9.525		11.97		21.7
	9.54		15.115		21.735
	9.57		15.125	Sines	6.01
	9.675		15.145		6.015
	9.755		15.17		6.115
	11.8		15.215		7.285
	11.815		15.225		7.295
	11.84		15.255		9.565
	11.98		15.29		9.585
	15.12		15.34		9.605
	15.195		15.355		9.61
	15.275		15.37		9.615
	17.81		15.38		9.635
	17.865		15.445		9.65
			17.725		9.66
PORTUGAL			17.735		9.665
Lisbon	5.97		17.75		9.67
	5.985		17.76		9.675
	6.07		17.77		9.685
	6.105		17.805		9.715
	6.115		17.835		11.825
	6.135		17.865		11.865
	7.115		17.875		11.905
	7.145		21.455		11.915
	7.165		21.51		15.16
	7.19		21.66		15.185
	7.2		21.665		15.245
	7.215		21.72		15.305
	7.245		21.73		15.315
	7.255		21.745		15.32
	7.295		21.75		17.82

	MHz		MHz		MHz
Portugal − contd		Bucharest	11.84	SAUDI ARABIA	
Sines	21.55		11.86	Diriyya	6.0
			11.885		7.22
QATAR			11.94		9.72
Doha	9.57		11.95		11.95
			11.97	Jeddah	9.67
REUNION			11.98		11.855
S Denis	7.245		15.25		15.115
			15.255	Riyadh	2.739
RHODESIA			15.335		3.97
Gwelo	3.307		15.34		5.26
	3.396		15.345		5.39
	4.828		15.365		5.876
	5.016		15.38		6.08
	5.975		15.39		6.085
	6.02		17.72		6.807
	7.175		17.73		7.11
	7.285		17.75		9.605
			17.785		9.63
ROUMANIA			17.79		9.73
Bucharest	5.99		17.805		9.887
	6.15		17.815		11.5
	6.155		17.825		11.685
	6.18		17.83		11.715
	6.19		17.835		11.72
	7.12		17.84		11.73
	7.175		17.85		11.775
	7.195		17.87		11.8
	7.225		17.89		11.82
	9.51		21.49		11.845
	9.53				11.855
	9.54	RWANDA			11.865
	9.55	Kigali	3.33		11.87
	9.57		6.055		11.88
	9.59		6.16		11.885
	9.595		7.225		11.89
	9.625		9.565		11.9
	9.64		9.7		11.905
	9.685		9.735		11.91
	9.69		11.765		11.915
	9.75		11.785		11.965
	11.705		11.965		12.113
	11.71		15.41		13.396
	11.725		17.765		15.06
	11.735		17.8		15.11
	11.74		21.54		15.115
	11.775	ST THOMAS			15.155
	11.79	ISLANDS			15.175
	11.81	Sao Thome	4.807		15.19
	11.83		5.339		15.22

	MHz		MHz		MHz
Saudi Arabia — contd		Mahe	15.406	Unknown	11.85
Riyadh	15.23		15.43		
	15.33		17.715	**SOLOMON ISLANDS**	
	15.35			Honiara	9.545
	15.365	**SIERRA LEONE**			
	15.37	Freetown	3.316	**SOMALIA**	
	15.665	Goderich	5.98	Hargeisa	7.16
	17.755	Waterloo	5.98		11.645
	17.76			Mogadiscio	6.095
	17.77	**SINGAPORE**			7.12
	21.48	Jurong	4.97		9.585
	21.5		5.01		
	21.505		5.052	**SOUTH AFRICA**	
	21.52		6.2	Bloemendal	3.25
	21.53	Kranji	6.01		3.388
	21.55		6.05		3.955
	21.555		6.065		3.965
	21.59		6.08		3.98
Unknown	10.385		6.195		3.995
			7.18		4.88
SENEGAL			9.57	Bloemfontein	3.32
Dakar	4.89		9.58		4.81
	7.17		9.59		4.875
	7.21		9.725	Johannesburg	3.23
	11.895		9.74	Meyerton	3.285
Tambacounda	6.045		11.71		4.835
	6.08		11.75		4.99
	6.18		11.85		5.98
Ziguinchor	3.336		11.865		6.01
	6.18		11.91		7.17
	7.21		11.92		7.27
			11.925		9.525
SEYCHELLES			11.955		9.56
Mahe	6.185		15.27		9.585
	9.545		15.28		9.61
	9.615		15.31		9.65
	9.65		15.435		9.68
	11.705		15.36		9.71
	11.725		15.38		11.79
	11.755		17.715		11.8
	11.785		17.88		11.885
	11.8	Singapore	6.0		11.9
	11.805		6.155		11.935
	11.84		7.17		11.955
	11.855		7.25		15.125
	11.86		9.53		15.155
	11.865		9.635		15.2
	11.945		11.94		15.22
	15.16		15.2		15.285
	15.325				

	MHz		MHz		MHz
South Africa — contd		Noblejas	6.12	Noblejas	17.865
Meyerton	15.38		6.19		17.87
	17.78		7.105		17.89
	17.805		7.185		21.535
	21.535		7.195		21.62
	21.555		9.505	Playa de Pals	7.155
	21.695		9.51		7.165
	25.79		9.52		7.19
			9.53		7.22
SPAIN			9.55		7.245
Arganda	5.97		9.56		7.295
	5.99		9.57		9.52
	6.0		9.58		9.625
	6.045		9.595		9.66
	6.065		9.6		9.68
	6.07		9.63		9.705
	6.1		9.69		9.715
	6.14		9.695		11.875
	6.16		9.725		11.885
	7.105		9.745		11.895
	7.11		11.715		11.915
	7.155		11.73		11.925
	7.2		11.74		11.935
	7.225		11.775		11.945
	7.245		11.785		11.97
	7.275		11.81		15.115
	9.505		11.815		15.13
	9.57		11.84		15.29
	9.68		11.85		15.34
	9.675		11.88		15.37
	9.685		11.89		15.38
	11.785		11.91		15.445
	11.83		11.92		17.725
	11.855		11.925		17.75
	11.92		11.93		17.76
	11.925		11.945		17.77
	11.93		11.965		17.865
	15.125		15.125		17.885
	15.205		15.185		17.895
	15.22		15.24		
	15.315		15.25		
	15.33		15.29	SUDAN	
	15.335		15.365	Khartoum	5.038
	15.395		15.375	Omdurman	6.15
Madrid	9.36		15.39		7.2
	9.79		15.395		9.505
Noblejas	6.05		15.405		11.835
	6.065		17.72		
	6.075		17.735	SURINAM	
	6.085		17.75	Paramaribo	4.85

	MHz		MHz		MHz
SWAZILAND		Hoerby	17.735	Schwarzen-	9.7
Manzini	5.955		17.77	burg	9.725
	6.07		17.79		9.765
	7.215		17.885		11.715
	7.255		21.505		11.72
	7.28		21.61		11.765
	9.52		21.615		11.77
	9.59		21.66		11.775
	9.73		21.69		11.78
	11.71	Karlsborg	6.065		11.782
	11.76		6.13		11.85
	11.81		7.145		11.87
	11.91		9.605		15.14
	11.955		9.63		15.17
	15.165		9.66		15.19
	17.745		9.665		15.23
Mbabane	3.223		9.69		15.235
	4.76		9.695		15.305
	4.79		11.705		15.35
			11.85		15.385
SWEDEN			11.905		15.43
Hoerby	6.065		15.24		17.715
	6.1		15.305		17.735
	9.59		17.735		17.74
	9.605		17.77		17.742
	9.635		21.65		17.77
	9.66		21.69		17.79
	9.665		21.7		17.83
	9.69	Varberg	11.952		17.84
	9.695		15.192		21.52
	9.745		17.777		21.545
	11.705		17.787		21.547
	11.79		21.552		21.57
	11.8		21.557		21.585
	11.81				21.63
	11.845	SWITZERLAND		Sottens	9.765
	11.85	Beromunster	7.21		11.72
	11.855	Geneva (UNO)	7.443		11.735
	11.905	Lenk	6.165		15.305
	11.915	Sarnen	9.535		15.43
	11.935	Schwarzen-	3.985		17.715
	11.955	burg	5.97		17.77
	15.12		6.045		17.815
	15.125		6.135		21.545
	15.13		7.21		21.57
	15.205		9.535	TAHITI	
	15.24		9.56	Papeete	6.135
	15.39		9.625		9.75
	15.415		9.66		11.825
	15.435		9.685		15.17

	MHz		MHz		MHz
TANZANIA		TURKEY		Kiev	7.39
Dar-es-Salaam	4.785	Ankara	6.185		9.58
	4.825		6.9		9.67
	5.05		7.17		9.71
	5.985		7.27		9.715
	6.105		9.515		9.72
	7.165		9.665		11.69
	7.23		11.8		11.77
	7.28		11.955		11.8
	9.53		11.965		11.925
	9.55		15.125		12.07
	9.595		15.135		15.11
	9.75		15.145		15.24
	15.435	Hakkari	7.649		17.82
Zanzibar	3.339	Unknown	6.34		17.86
					21.515
THAILAND		UGANDA		Lvov	6.115
Bangkok	4.832	Kampala	3.334		6.125
	6.07		4.976		7.12
	6.08		5.026		7.15
	7.115		6.03		7.16
	9.655		9.515		7.23
	11.905		9.685		9.685
			9.73		9.705
TIBET			15.235		9.735
Lhasa	4.035		15.25		9.765
	5.935		15.325		11.72
	9.49				11.75
	9.655	UKRAINE			11.78
		Ivanofransk	7.115		11.9
TIMOR			9.635		15.135
Dili	3.55		9.665		15.175
	3.85		9.76		17.76
			11.88		17.78
TOGO			11.955		17.845
Lama Karma	3.222		15.33		17.89
Lome	5.047		17.785		21.565
Togblekope	7.265	Kharkov	11.76	Simferopol	6.03
			11.95		6.04
TRISTAN DA CUNHA			11.97		6.09
Tristan da-Cunha	3.29	Kiev	4.92		6.13
			4.94		7.105
			5.915		7.195
TUNIS			6.02		9.515
Sfax	7.225		6.085		9.535
	7.275		6.14		9.565
	11.97		6.165		9.675
	15.225		7.165		9.74
			7.23		9.75
			7.3		11.705

	MHz		MHz		MHz
Ukraine — contd		London	5.955	London	7.285
Simferopol	11.81		5.965		7.295
	15.13		5.975		7.32
	15.185		5.99		7.325
	15.205		5.995		7.844
	15.245		6.005		7.848
	15.26		6.01		7.893
	15.325		6.015		7.973
	15.425		6.03		7.976
	17.815		6.04		7.991
	17.855		6.05		9.317
	21.515		6.06		9.323
	21.745		6.065		9.41
Starobelsk	6.055		6.07		9.505
	6.145		6.08		9.51
	6.17		6.085		9.515
	7.175		6.11		9.53
	9.58		6.115		9.54
	9.59		6.12		9.555
	15.33		6.125		9.565
Vinnitsa	6.1		6.13		9.57
	6.175		6.14		9.58
	7.215		6.15		9.6
	7.245		6.16		9.615
	9.53		6.17		9.62
	9.575		6.18		9.625
	9.665		6.185		9.635
	9.72		6.195		9.64
	11.73		6.838		9.645
	11.735		7.105		9.66
	11.85		7.11		9.68
	15.135		7.12		9.69
	15.17		7.14		9.7
	15.435		7.15		9.705
	17.755		7.155		9.715
	17.775		7.17		9.735
	17.835		7.18		9.745
	17.885		7.185		9.75
			7.2		9.76
UNITED ARAB			7.21		9.77
EMIRATES			7.22		9.825
Abu Dhabi	9.62		7.23		9.858
Dubai	6.04		7.235		9.915
			7.24		11.68
UNITED KINGDOM			7.245		11.71
London	3.952		7.25		11.715
	3.97		7.255		11.72
	3.975		7.26		11.74
	4.467		7.265		11.75
	5.872		7.28		11.76

	MHz		MHz		MHz
United Kingdom –		London	15.435	Bethany	16.222
contd	11.775		15.44		19.261
London	11.78		15.445		25.88
	11.785		15.589	Boulder Std	5.0
	11.79		15.67	Freq.	10.0
	11.805		15.849		15.0
	11.83		15.91		20.0
	11.835		17.642	Cincinnati	6.03
	11.845		17.695		6.04
	11.85		17.705		6.125
	11.855		17.715		6.155
	11.865		17.74		9.525
	11.905		17.76		9.685
	11.91		17.77		9.755
	11.915		17.78		11.74
	11.925		17.79		11.79
	11.935		17.795		11.83
	11.945		17.8		11.835
	11.955		17.81		11.895
	11.96		17.815		15.14
	12.04		17.855		15.16
	12.095		17.86		15.19
	12.127		17.87		15.25
	12.179		18.08		15.33
	15.07		19.455		15.43
	15.105		20.345		17.71
	15.11		21.47		21.47
	15.12		21.5		21.485
	15.14		21.55		21.5
	15.16		21.57		21.66
	15.165		21.59	Delano	6.01
	15.18		21.61		6.145
	15.195		21.63		6.185
	15.205		21.64		9.545
	15.215		21.65		9.565
	15.235		21.695		9.65
	15.245		21.71		9.66
	15.25		22.93		9.68
	15.26		23.191		11.82
	15.265	Rugby Stan-	2.5		11.83
	15.27	dard	5.0		11.85
	15.28	Frequency	10.0		11.865
	15.31				11.9
	15.315	UPPER VOLTA			11.915
	15.32	Ouagadougou	4.815		13.86
	15.35		7.23		15.205
	15.355				15.225
	15.39	U.S.A.			15.27
	15.42	AFRTS	9.929		15.345
	15.425	Bethany	10.869		15.365

213

Location	MHz
U.S.A. — contd	
Delano	17.72
	17.75
	17.78
	17.84
	17.895
	19.912
	21.46
	21.695
	21.745
	25.99
Dixon	5.955
	6.015
	6.055
	6.095
	6.125
	6.185
	9.545
	9.65
	9.66
	9.7
	9.73
	11.74
	11.805
	11.83
	11.85
	11.9
	11.92
	15.25
	15.33
	15.345
	17.72
	17.75
	17.765
	17.82
	17.84
	18.135
	19.48
	21.61
	21.65
	21.745
	25.62
	26.095
Greenville	5.745
	5.995
	6.02
	6.03
	6.055
	6.08
	6.125
Greenville	6.13
	6.135
	6.19
	6.873
	7.651
	7.768
	7.77
	9.51
	9.53
	9.54
	9.565
	9.58
	9.6
	9.62
	9.625
	9.635
	9.64
	9.65
	9.67
	9.68
	9.715
	9.725
	9.745
	9.765
	10.235
	10.38
	10.454
	10.869
	11.715
	11.73
	11.74
	11.75
	11.79
	11.805
	11.83
	11.835
	11.84
	11.845
	11.855
	11.89
	11.9
	11.905
	11.915
	11.92
	11.935
	11.955
	11.96
	11.965
	13.491
	15.12
Greenville	15.125
	15.14
	15.15
	15.155
	15.16
	15.17
	15.195
	15.205
	15.225
	15.235
	15.24
	15.25
	15.26
	15.28
	15.315
	15.35
	15.375
	15.395
	15.4
	15.41
	15.415
	15.425
	15.43
	15.65
	15.752
	15.77
	16.43
	17.73
	17.74
	17.755
	17.775
	17.785
	17.8
	17.815
	17.82
	17.855
	17.86
	17.895
	18.275
	18.782
	19.505
	19.721
	20.06
	20.125
	21.47
	21.5
	21.535
	21.54
	21.55
	21.59

	MHz
U.S.A. — contd	
Greenville	21.61
	21.67
	21.69
	26.04
New York	13.272
Okeechobee	9.525
	9.555
	9.63
	9.66
	9.69
	9.715
	11.705
	11.72
	11.77
	11.8
	11.805
	11.815
	11.845
	11.85
	11.855
	11.875
	11.885
	11.925
	15.11
	15.115
	15.13
	15.16
	15.195
	15.215
	15.44
	17.73
	17.785
	17.805
	17.845
	17.87
	17.875
	21.525
	21.615
Red Lion	11.905
	15.145
	15.175
	15.185
	15.305
	17.72
	17.73
	21.565
Redwood City	5.98
	9.505
	9.53

	MHz
Redwood City	9.555
	9.6
	9.615
	9.625
	9.685
	9.69
	9.75
	15.28
	15.355
Scituate	5.985
	6.155
	9.525
	9.69
	9.715
	11.78
	11.805
	11.815
	11.85
	11.855
	11.925
	15.11
	15.13
	15.135
	15.16
	15.26
	15.27
	15.35
	15.39
	15.44
	17.73
	17.785
	17.845
	17.865
	17.875
	21.525
	21.61
Unknown	5.23
	5.703
	10.537
U.S.S.R.	
Achkhabad	4.825
	4.895
	6.125
	11.93
	15.195
	17.74
Alma Ata	4.545
	4.99
	5.26

	MHz
Alma Ata	5.915
	5.96
	5.97
	6.06
	6.135
	6.18
	6.87
	7.14
	7.185
	9.505
	9.61
	11.745
	11.92
	11.95
	15.11
	15.155
	15.21
	15.23
	17.85
Arkhangelsk	5.015
Armavir	5.915
	5.96
	5.965
	6.065
	6.105
	6.12
	6.13
	6.17
	7.155
	7.17
	7.19
	7.205
	7.215
	7.265
	9.52
	9.575
	9.61
	9.62
	9.655
	9.675
	9.765
	11.71
	11.745
	11.775
	11.88
	11.9
	11.93
	11.96
	15.155
	15.18

Station	MHz	Station	MHz	Station	MHz
U.S.S.R. — contd		Duchanbe	7.115	Frunze	17.745
Armavir	15.21		7.275		17.775
	15.24		9.605		17.785
	15.265		9.735		17.815
	15.28		9.755		17.825
	15.32		11.715		17.84
	15.35		11.825		21.515
	15.375		11.9		21.53
	15.385		15.155		21.625
	15.395		15.22		21.745
	15.405		17.765	Gorkii	7.175
	17.73		21.505		7.185
	17.775		21.585		11.86
	17.84	Dzhambul	4.76		15.265
	17.85	Erevan	4.04		15.385
	17.86		4.81	Irkutsk	5.98
	21.51		4.99		6.05
	21.645		6.055		6.09
	21.53		7.27		6.24
	21.555		9.515		7.2
	21.635		9.685		7.265
	21.705		11.72		9.51
	21.745		11.76		9.555
Baku	4.785		11.89		9.565
	4.957		12.04		9.575
	6.11		15.15		9.59
	6.135		15.19		9.67
	6.195		15.435		9.685
	9.65		17.735		9.735
	11.92		17.755		9.755
	15.24		21.505		11.765
	15.26	Frunze	4.01		11.775
	15.415		6.08		15.175
	17.8		7.205		15.195
	17.825		7.23		15.21
	21.49		9.53		15.285
	21.68		9.565		15.36
Blagovesh-	4.975		9.655		15.42
chensk			9.695		17.73
Blagoevechtchen			9.71		17.805
	5.96		11.71	Iujnsakh-	7.26
	7.295		11.72	klinsk	11.86
	9.52		11.74	Jigulevsk	6.17
	9.58		11.82		7.2
	11.79		11.85		11.895
	11.94		11.89		15.115
	15.17		11.905		15.185
Duchanbe	4.635		11.97		15.375
	6.105		15.265	Kalatch	15.2
	6.12		15.435	Kalinin	6.105

	MHz		MHz		MHz
U.S.S.R. — contd		Kenga	7.275	Khabarovsk	17.87
Kalinin	7.31		9.505		21.505
	21.635		9.555	Khanty-	4.52
Kaunas	6.1		9.6	Mansiysk	
	9.635		9.7	Kinghisepp	7.175
	9.71		9.745		9.665
	11.87		9.77		9.76
Kazan	6.04		11.73	Komsomols-	5.97
	6.065		11.795	kamur	6.03
	6.11		11.87		6.06
	6.15		11.89		6.08
	6.175		11.96		6.15
	7.125		15.175		7.215
	7.14		15.255		7.265
	7.16		15.265		7.275
	7.17		15.375		7.28
	7.18		15.425		9.505
	7.2	Khabarovsk	5.985		9.53
	7.21		6.02		9.71
	7.23		6.07		9.75
	7.285		6.115		11.71
	9.51		6.135		11.715
	9.52		6.145		11.81
	9.58		6.175		11.9
	9.59		7.11		11.91
	9.695		7.125		11.97
	9.7		7.175		15.35
	9.795		7.21		17.865
	11.71		7.245	Konevo	11.85
	11.735		9.545	Krasnoiarsk	5.29
	11.75		9.61		6.01
	11.765		9.645		6.06
	11.805		9.7		6.125
	11.845		11.705		6.155
	11.85		11.72		6.175
	11.92		11.73		7.15
	11.96		11.82		7.23
	11.965		11.87		7.245
	15.175		15.115		7.275
	15.26		15.21		7.285
	15.295		15.255		9.56
	15.32		15.28		9.75
	15.375		15.295		11.76
	15.4		15.405		11.78
	15.415		15.42		11.82
	17.72		17.72		11.835
	17.82		17.775		11.86
Kenga	7.125		17.805		11.925
	7.255		17.85		11.94
	7.27		17.86		11.96

Location	MHz	Location	MHz	Location	MHz
U.S.S.R. — contd		Moskva	6.16	Moskva	9.85
Krasnoiarsk	15.11		6.175		10.12
	15.28		6.2		10.338
	15.42		6.945		10.74
	17.775		7.19		11.34
	17.79		7.205		11.63
	17.835		7.21		11.69
Kursk	6.1		7.27		11.75
	15.33		7.28		11.795
	17.745		7.285		11.83
	17.86		7.29		11.845
	21.54		7.295		11.88
Leningrad	5.95		7.3		11.91
	6.2		7.325		11.92
	7.305		7.34		11.93
	9.59		7.35		11.975
	9.765		7.355		11.98
	11.755		7.36		11.985
	11.765		7.38		12.045
	15.18		7.395		12.055
	15.375		7.4		12.07
	21.6		7.41		15.1
Magadan	4.03		7.42		15.21
	5.94		7.43		15.36
	7.32		7.44		15.375
	12.24		9.13		15.41
Minsk	5.945		9.15		15.455
	7.33		9.45		17.71
	7.37		9.48		17.835
	7.42		9.49		17.9
	9.795		9.5		21.45
	12.005		9.53		21.575
Moskva	4.825		9.54	Murmansk	5.93
	4.85		9.55	Nikolaevvs-	5.97
	4.86		9.6	kamur	5.99
	5.77		9.62		6.09
	5.9		9.625		6.155
	5.905		9.63		6.19
	5.91		9.64		7.23
	5.94		9.65		9.725
	5.965		9.685		9.735
	5.97		9.74		9.755
	6.0		9.745		11.79
	6.01		9.765		15.24
	6.03		9.775		15.245
	6.045		9.78		15.415
	6.08		9.785		17.705
	6.12		9.79		17.765
	6.13		9.8		17.845
	6.145		9.81	Novosibirsk	6.045

	MHz		MHz		MHz
U.S.S.R. — contd			17.88	Sverdlovsk	7.21
Novosibirsk	6.12	Petrozavodsk	4.78		7.235
	6.13		5.065		7.25
	6.185	Riazan	5.98		9.75
	7.02		6.185		11.715
	7.12		7.265		11.76
	7.145		9.72		11.785
	7.17		11.79		11.91
	7.185		11.89		11.94
	9.645		15.14		15.305
	9.675		15.275		15.33
	9.735		15.44		15.405
	11.74		17.775		15.505
	11.865		17.87		17.845
	11.885	Riga	5.935		17.89
	11.92		7.14	Tashkent	4.85
	15.24		7.235		5.035
	15.3		15.14		5.925
	15.31		15.22		5.945
	15.335		21.615		5.975
	15.385	Serpukhov	5.99		6.025
	15.4		6.065		7.145
	15.445		6.095		7.17
Okhotsk	9.6		6.15		7.22
Omsk	6.19		7.15		9.52
	7.11		7.165		9.54
	7.16		9.51		9.55
	9.59		9.56		9.6
	11.77		9.605		9.66
	15.15		9.63		9.75
Orcha	11.985		9.73		11.73
Orenburg	6.09		9.77		11.835
Palana	4.52		11.705		11.925
Petropavlovsk	4.055		11.76		11.96
	4.485		11.78		15.115
	5.94		11.81		15.255
Petropavlo Kam	6.1		11.835		15.265
	6.185		11.845		15.33
	7.16		11.87		15.395
	7.29		11.96		17.71
	9.54		15.15		17.755
	9.695		15.185		17.86
	9.71		15.42		17.87
	9.75		15.45		17.875
	11.915		17.73		21.585
	15.14		17.795		21.725
	15.18		21.59	Tallin	6.055
	15.21	Sverdlovsk	5.92		6.085
	15.425		5.96		17.805
	17.72		6.075	Tbilisi	5.04

	MHz		MHz		MHz
U.S.S.R. — contd		Tula	6.18	Vladivostok	5.015
Tbilisi	5.93		7.12		5.96
	5.98		7.145		6.035
	6.06		7.16		6.165
	6.085		7.195		7.2
	7.115		7.24		7.22
	7.185		7.25		7.32
	7.255		7.285		7.49
	9.625		9.545		9.51
	9.695		9.555		9.53
	9.76		9.61		9.62
	11.71		9.69		9.635
	11.755		9.7		9.71
	11.78		9.73		9.75
	11.795		9.735		9.77
	11.805		11.72		9.79
	11.82		11.935		11.755
	11.96		11.95		15.13
	15.305		15.17		17.745
	17.805		15.19		17.755
	17.885		15.37		17.86
Tchita	4.86		15.435	Volgograd	6.075
	5.97		17.765		7.265
	6.0		17.805	Vologda	5.91
	6.045		21.46		11.725
	6.055		21.49		15.35
	7.115	Tyumen	4.895	Voronej	6.005
	7.15	Ufa	4.485		6.14
	7.22	Ulan Ude	4.795		9.665
	7.315	U.S.S.R. Freq. Standard	14.996		9.675
	9.575	U.S.S.R. Met. Station	3.417		11.82
	9.625		3.44		15.23
	9.665		3.47		15.295
	9.68		3.93		15.305
	9.7		4.647	Yakutsk	4.395
	11.755		4.656	Yuzhno Sakhalinsk	4.05
	11.81		4.665	Unknown	3.46
	11.88		4.678		3.94
	11.89		4.685		3.99
	11.925		4.695		4.425
	15.19		6.56		4.645
	15.385		6.596		4.654
	15.42		6.615		4.71
	17.815		6.617		4.72
	17.89		6.73		4.73
Tula	5.905	Vladivostok	3.96		4.747
	5.97		3.995		4.765
	5.995		4.04		4.815
	6.07		4.465		4.82
	6.075				

	MHz		MHz		MHz
U.S.S.R. — contd		Unknown	7.37	Unknown	12.1
Unknown	4.83		7.385		12.14
	4.835		7.437		12.165
	4.84		7.53		12.175
	4.93		7.54		12.205
	4.955		7.55		12.222
	5.255		7.605		12.25
	5.29		7.67		12.252
	5.32		7.74		12.28
	5.455		7.925		13.37
	5.47		7.948		13.38
	5.7		8.125		13.59
	5.794		8.903		13.71
	5.815		8.91		13.76
	5.83		8.917		13.82
	5.885		8.97		13.96
	5.92		9.2		14.29
	5.935		9.21		14.44
	5.95		9.24		14.595
	6.205		9.34		14.85
	6.215		9.345		15.045
	6.23		9.39		15.06
	6.243		9.44		15.46
	6.285		9.46		15.465
	6.39		9.47		15.48
	6.4		9.57		15.49
	6.548		9.595		15.497
	6.554		9.815		15.5
	6.568		9.82		15.515
	6.77		9.83		15.52
	6.808		10.09		15.525
	6.822		10.62		15.53
	6.825		10.66		15.535
	6.852		10.69		15.54
	6.89		10.695		15.545
	6.905		10.855		15.6
	6.92		11.495		15.78
	6.98		11.575		15.85
	6.987		11.67		15.87
	7.01		11.7		16.03
	7.03		11.99		16.14
	7.04		12.0		16.19
	7.045		12.01		16.25
	7.08		12.02		16.33
	7.1		12.025		16.87
	7.105		12.03		17.135
	7.305		12.035		17.56
	7.31		12.05		17.58
	7.335		12.06		17.69
	7.345		12.075		17.7

U.S.S.R. — contd

Location	MHz
Unknown	18.015
	18.195
	18.285
	18.31
	18.37
	18.46
	18.653
	18.83
	19.13
	19.21
	19.725
	19.833
	19.845
	20.06
	20.25
	20.605
	21.495
	21.665
	22.205
	22.77

VATICAN

Location	MHz
S M Galeria	5.995
	6.015
	6.19
	7.155
	7.16
	7.25
	7.235
	9.55
	9.605
	9.615
	9.625
	9.645
	9.73
	11.705
	11.715
	11.725
	11.74
	11.765
	11.81
	11.82
	11.83
	11.845
	11.855
	11.875
	15.12
	15.165
	15.2
S M Galeria	15.21
	15.4
	17.705
	17.785
	17.825
	17.84
	17.845
	17.895
	21.485
Vatican City	6.105
	6.19
	6.21
	6.221
	6.23
	7.04
	11.7
	17.79
	17.9

VENEZUELA

Location	MHz
Barcelona	3.385
Barquesimeto	4.8
	4.82
	4.88
	4.9
	4.99
Bolivar	4.77
Caracas	3.245
	4.87
	4.89
	4.92
	5.02
	5.03
	5.05
	6.1
	6.17
	9.64
	9.66
	11.75
	15.39
Carora	4.91
Cumana	4.96
El Tigre	3.25
Maracaibo	4.81
	4.86
Merida	3.395
Puerto Cabello	3.285
San Cristobal	4.93
	4.98
San Filipe	4.94
Tovar	3.225
	3.25
Trujillo	3.295
Valencia	3.355
	4.78
Valera	4.84
Villa de Cura	4.97
Unknown	4.96

VIETNAM

Location	MHz
Bac Thai	6.884
	7.08
Cao Lang	4.785
	6.26
Gia Lai-Cong Tum	4.762
Ha Bac	4.647
Ha Tuyen	4.82
Hanoi	4.0
	4.86
	4.932
	4.94
	4.945
	4.995
	6.426
	6.435
	7.375
	7.385
	7.415
	7.47
	7.512
	9.84
	9.887
	10.01
	10.04
	10.06
	10.225
	12.033
	15.009
	15.02
Hoang Wen Sun	5.594
Cho Chi Min City	6.165
	9.62
	9.625
Hue	4.68
Lai Chau	5.927
Nghe Tinh	6.28
Phu Kmanh	5.139
Quang Minh	4.692

	MHz		MHz		MHz
Vietnam — contd		Lusaka	7.235	RFE/R	15.73
Son La	4.77		7.25	Liberty	15.775
	6.331		9.58		16.065
Thanh Hoa	4.881		11.88		16.24
Thu Dau Mot	4.6		17.895		17.445
City					20.215
Vinh Phu	4.284	ZANZIBAR			20.71
		Marhubi	6.005		22.97
YEMEN					25.35
Sanaa	4.853	LOCATIONS		R Free	14.44
	4.975	UNKNOWN		Portugal	
	6.05	Bizam R.	5.914	R Freedom	
	7.265		6.2	for South	
	9.77		9.5	Yemen	9.953
	9.78		9.585	R Kulmis	7.235
		German Figure		R of the	6.1
YUGOSLAVIA		Groups	6.453	Patriots	9.5
Belgrade	6.1	Kashmir R.	3.915		9.65
	6.15	Mebo II	6.206		11.7
	7.2	National Voice		R. Sandino	7.588
	7.24	of Iran	6.025	R. Yemen A	
	9.505	R Echo of		Arab Rep	6.005
	11.735	Hope	6.348	V of Arab	
	15.24	RFE/R Liberty	4.475	Syria	9.51
			4.505	V of Arabian	
ZAIRE			4.565	Peninsula	
Badundu	7.115		4.695	People	8.36
Kananga	6.125		5.125	V of Black Man's	
Kinshasa	7.255		5.295	Resistance	7.2
	9.66		5.79	V of Commun-	
	9.77		5.845	ist Party of	
	11.72		5.945	Turkey	6.2
	11.795		5.98	V of Demo-	4.908
	15.245		6.97	cratic	6.09
	15.35		6.975	Kampuchea	6.825
Kisangani	6.085		6.995		7.35
Lubumbashi	4.75		7.44		9.47
	7.205		9.09		9.755
	11.865		9.091		11.6
Mbandaka	5.995		9.17		11.725
Mbujimayi	7.295		9.25		11.99
			10.19		15.27
ZAMBIA			10.315		15.6
Lusaka	3.295		10.42		17.705
	3.346		11.575	V of Eritrean	
	4.91		11.675	Revolution	7.24
	4.965		12.246	V of Front	
	6.06		13.69	for Redemp-	
	6.165		14.712	tion of	
	7.22		14.715	Somalia	9.59

	MHz		MHz		MHz
Locations Unknown — contd		V of the Namibian People	7.18	V of the People of Thailand	9.42
V of Iraqi Kurdistan	5.95	V of NUFK	4.675	V of the Rev Party for Reunification	4.557
	5.99		7.015		
	6.94		9.985		
	10.055		10.08		
V of Kawthulay	4.88		10.12		
V of Lebanon	6.852		12.006		
V of the One Lebanon	9.51	V of Peace	6.25		
V of Malayan Revolution	7.305	V of the People	7.103		
	9.62	V of the People of Burma	5.110		
	11.83				
	15.79				

EUROPEAN
V.H.F. SOUND BROADCASTING STATIONS

This list includes only those transmitters in Europe with an e.r.p. of 100 kW or more, except in the case of the U.K. where all stations are listed. There are in addition more than 3,500 lower powered transmitters in Europe of which over 1,600 are in Italy. The carrier frequencies of the channel numbers in the first column are given on p. 230. Some carrier frequencies are offset from that allocated by up to 150 kHz. Stations transmitting stereophonic programmes are marked with an 'S'.

		kW
AUSTRIA		
6	Lichtenberg (S)	100
7	Schoeckl (S)	100
8	Jauerling (S)	100
9	Pfaender (S)	100
10	Kahlenberg	100
13	Gaisberg (S)	100
14	Schoeckl (S)	100
15	Jauerling	100
16	Kahlenberg (S)	100
19	Dobratsch·Villacher (S)	100
21	Pfaender (S)	100
27	Lichtenberg	100
28	Schoeckl	100
33	Jauerling (S)	100
35	Lichtenberg (S)	100
36	Dobratsch-Villacher	100
36	Kahlenberg	100
37	Pfaender	100
43	Kahlenberg (S)	100
47	Dobratsch-Villacher (S)	100
EIRE		
9	Truskmore (S)	120
22	Mullaghanish (S)	120
24	Maghera (S)	120
26	Mount Leinster (S)	120
FRANCE		
4	Carcassonne	125
6	Lille (S)	150
	Paris	100
7	Le Mans	100
	Reims (S)	150
8	Limoges	150

		kW
10	Rennes (S)	100
12	Nantes	200
13	Carcassonne (S)	125
14	Niort	200
17	Rouen (S)	100
19	Le Mans	100
20	Limoges	150
21	Paris (S)	100
22	Rennes	100
23	Rouen	100
24	Nantes	200
26	Lille	150
31	Niort	200
32	Carcassonne	125
	Rouen	100
33	Reims	150
	Le Mans (S)	100
35	Limoges (S)	150
	Paris	100
37	Lille	150
38	Rennes	100
39	Reims	150
40	Nantes (S)	200
41	Niort	200
GERMANY (East)		
9	Karl Marx Stadt	100
11	Inselberg	100
13	Marlow	100
15	Berlin (S)	100
	Brocken	100
16	Sonneberg	100
18	Inselberg (S)	100
23	Leipzig	100
24	Sonneberg	100
27	Berlin	100
	Schwerin	100

Germany (East) — contd	kW
29 Berlin (S)	100
33 Karl Marx Stadt	100
34 Inselberg	100
35 Brocken	100
Berlin	100
38 Schwerin	100
42 Berlin	100

GERMANY (West)	
3 Gottelbörner Hoehe (S)	100
4 Bremen (S)	100
6 Gruenten/Allgaeu (S)	100
Heidelberg (S)	100
Langenberg	100
8 Wendelstein (S)	100
10 Harz	100
10 Stuttgart/Degerloch (S)	100
12 Gruenten/Allgaeu (S)	100
Ochsenkopf (S)	100
Teutoburger Wald	100
14 Gottelborner Hoehe (S)	100
17 Brotjacklriegel (S)	100
Harz (S)	100
Stuttgart/Degerloch	100
20 Kreuzberg/Rhoen (S)	100
21 Teutoburger Wald (S)	100
22 Wendelstein (S)	100
23 Bremen (S)	100
Waldenburg (S)	100
25 Brotjacklriegel	100
26 Stuttgart/Degerloch (S)	100
27 Langenberg (S)	100
28 Gottelborner Hoehe (S)	100
29 Gruenten/Allgaeu	100
30 Ochsenkopf (S)	100
31 Kreuzberg/Rhoen	100
32 Brotjacklriegel (S)	100
Waldenburg	100
33 Teutoburger Wald (S)	100
36 Heidelberg (S)	100
37 Harz (S)	100
38 Kreuzberg/Rhoen (S)	100
Wendelstein	100
39 Waldenburg (S)	100

	kW
41 Langenberg (S)	100
Ochsenkopf	100
43 Heidelberg	100
51 Stuttgart-Frauenkapf (S)	100

LUXEMBOURG	
6 Hosingen	100
33 Marnach	100

NETHERLANDS	
4 Roermond (S)	100
13 Roermond (S)	100
25 Roermond (S)	100

NORWAY	
5 Oslo	100
22 Bokn	100

UNITED KINGDOM	
4 Ballachulish	15W
Betws-y-Coed	10W
Bressay	10
Ffestiniog	50W
Forfar(S)	10
Llanidloes (S)	5W
4 Lochgilphead (S)	10W
Londonderry	13
North Hessary Tor (S)	60
Sandale (S)	120
Sutton Coldfield (S)	120
Wensleydale (S)	25W
5 Barnstaple (S)	150W
Campbeltown (S)	35W
Carmel (S)	3.2
Douglas	6
Newry	30W
Pontop Pike (S)	60
Rowridge (S)	60
R. Sheffield†	30W
Skriaig	10
Toward (S)	250W
Windermere (S)	20W
6 Ayr (S)	55W
Bath (S)	35W
Belmont (S)	8
Blaen-Plwyf	60
Brecon	10W
Brougher Mountain (S)	2.5
Cambridge	20W

		kW
14	Northampton (S)	60W
	Oban	1.5
	Perth	15W
	Wrotham (S)	120
15	Fort William	1.5
	Haverfordwest	10
	Holme Moss (S)	120
	Machynlleth	60W
	Orkney	20
	Pitlochry	200W
	Rosneath (S)	25W
	Ventnor (S)	20W
16	Hereford (S)	25W
	Kilvey Hill (S)	1
	Kinlochleven	2W
	Llanddona	12
	Oxford (S)	22
	Penifiler	6W
	Redruth (S)	9
	Rosemarkie	12
	Tacolneston (S)	120
	Weardale (S)	100W
	Whitby (S)	40W
17	Grantown	350W
	Kirk O'Shotts (S)	120
	Morecambe Bay (S)	4
	Scarborough (S)	25W
	Sheffield (S)	60W
	Wenvoe (S)	120
18	Ballachulish	15W
	Brighton (S)	150W
	Chatton (S)	5.6
	Divis (S)	60
	Dolgellau	15W
	Ffestiniog	50W
	R. Leeds†	5.2
	Llanidloes (S)	5W
	North Hessary Tor	60
	Peterborough	20
	Sandale	120
	Swingate (S)	7
	Thrumster	10
19	Betws-y-Coed	10W
	Bressay	10
	Campbeltown	35W
	Carmel (S)	3.2
	Douglas	6
	Forfar (S)	10
	Lochgilphead	10W
19	Londonderry	13
19	Sutton Coldfield (S)	120
	Wensleydale (S)	25W
20	Ayr (S)	55W
	Barnstaple (S)	150W
	Belmont (S)	8
	Blaen-Plwyf	60
	Campbeltown	35W
	Carmarthen	10W
	Kendal (S)	25W
	Maddybenny More (S)	30W
	Meldrum	60
	Newry	30W
	Okehampton	15W
	Pontop Pike (S)	60
	Rowridge	60
	Skriaig	10
	Toward (S)	250W
	Windermere (S)	20W
21	Ballycastle (S)	40W
	Bath (S)	35W
	Brecon	10W
	Brougher Mountain (S)	2.5
	Carmarthen (S)	10W
	Cambridge	20W
	Churchdown Hill (S)	25W
	Isles of Scilly	20W
	Kilkeel (S)	25W
	Llangollen	10
	Millburn Muir (S)	25W
	Northampton (S)	60W
	Oban	1.5
	Perth	15W
22	Ashkirk (S)	18
	Fort William	1.5
	Haverfordwest	10
	Holme Moss (S)	120
	Kingussie	35W
	Larne (S)	15W
	Llandrindod Wells (S)	1.5
	Melvaig	22
	Orkney	20
	Pitlochry	200W
	Rosneath (S)	25W
	Wrotham (S)	120
23	Kilvey Hill (S)	1
	Llanddona	12
	Machynlleth	60W
	Oxford (S)	22
	Penifiler	6W
	Rosemarkie	12

United Kingdom — contd kW

		kW
23	Ventnor	20W
	Whitby (S)	40W
24	R. Derby†	10W
	Grantown	350W
	Hereford (S)	25W
	Kinlochleven	2W
	Kirk O'Shotts (S)	120
?	Redruth	9
	Scarborough (S)	25W
	Sheffield (S)	60W
	Tacolneston	120
	Weardale (S)	100W
	Wenvoe (S)	120
25	Brighton	150W
	Chatton (S)	5.6
	Divis (S)	60
	Dolgellau	15W
	Morecambe Bay (S)	4
	Peterborough	20
	Swingate (S)	7
	Thrumster	10
26	BRMB Radio (S)*	2
	Les Platons	1.5
	R. London†	16.5
	Sandale	120
27	R. Clyde*	3.4
	R. Hallam (S)*	100W
	R. Leicester†	300W
	R. Manchester†	4.2
	R. Oxford †	4.5
	Swansea Sound (S)*	1
	R. Tees (S)*	2
	R. Victory (S)*	200W

		kW
28	R. Brighton†	500W
	R. Bristol†	5
	R. Newcastle†	3.5
	R. Nottingham†	300W
29	R. Birmingham†	5.5
	Capital Radio*	2
	R. Carlisle†	5
	R. Merseyside†	7.5
30	Downtown Radio (S)*	1
	R. Hallam (S)*	50W
	Pennine Radio (S)*	500W
	Plymouth Sound (S)*	1
	R. Solent†	5
	R. Stoke-on-Trent†	2.5
31	R. Blackburn†	1.6
	R. Trent (S)*	300W
32	R. City (S)*	5
	R. Cleveland†	5
	R. Derby†	5.5
	R. Medway†	5.6
33	R. Forth*	500W
	R. Humberside†	4.5
	Metro Radio (S)*	5
	R. Piccadilly (S)*	2
	Thames Valley Broadcasting*	250W
	Wenvoe (S)	120
34	Beacon Radio (S)*	1
	LBC*	2
	Les Platons	1.5
	R. Orwell (S)*	1
35	R. Sheffield†	5.2

*Independent Local Radio
†BBC Local Radio

EUROPEAN
V.H.F. SOUND BROADCASTING CHANNELS

Channel	MHz	Channel	MHz
1	87.2–87.4	31	96.2–96.4
2	87.5–87.7	32	96.5–96.7
3	87.8–88.0	33	96.8–97.0
4	88.1–88.3	34	97.1–97.3
5	88.4–88.6	35	97.4–97.6
6	88.7–88.9	36	97.7–97.9
7	89.0–89.2	37	98.0–98.2
8	89.3–89.5	38	98.3–98.5
9	89.6–89.8	39	98.6–98.8
10	89.9–90.1	40	98.9–99.1
11	90.2–90.4	41	99.2–99.4
12	90.5–90.7	42	99.5–99.7
13	90.8–91.0	43	99.8–100.0
14	91.1–91.3	44	100.1–100.3
15	91.4–91.6	45	100.4–100.6
16	91.7–91.9	46	100.7–100.9
17	92.0–92.2	47	101.0–101.2
18	92.3–92.5	48	101.3–101.5
19	92.6–92.8	49	101.6–101.8
20	92.9–93.1	50	101.9–102.1
21	93.2–93.4	51	102.2–102.4
22	93.5–93.7	52	102.5–102.7
23	93.8–94.0	53	102.8–103.0
24	94.1–94.3	54	103.1–103.3
25	94.4–94.6	55	103.4–103.6
26	94.7–94.9	56	103.7–103.9
27	95.0–95.2	57	104.0–104.2
28	95.3–95.5	58	104.3–104.5
29	95.6–95.8	59	104.6–104.8
30	95.9–96.1	60	104.9–105.1